FUNDAMENTAL PRINCIPLES OF PHYSICS

FUNDAMENTAL

F. WOODBRIDGE CONSTANT, Trinity College *, 1904 -*

PRINCIPLES OF PHYSICS

ADDISON-WESLEY PUBLISHING COMPANY
READING, MASSACHUSETTS · PALO ALTO · LONDON · DON MILLS, ONTARIO

This book is in the

ADDISON-WESLEY SERIES IN PHYSICS

Preface

This is primarily a textbook for a one-semester liberal arts physics course at the college level; however, it could also serve as the basic text, accompanied by supplementary reading, for a more leisurely one-year course. This book is a shortened and less mathematical version of the author's *Fundamental Laws of Physics*, with less space devoted to classical mechanics and electromagnetism and more to the recent concepts of atomic and nuclear physics, elementary particles, symmetry and conservation principles, and the basic forces of nature.

The present text is concerned with the fundamental nature of physical science and the methods of scientific discovery. It is not an encyclopedia of all the interesting facts in physics and their engineering applications, but rather it follows historically the main line of physical discovery from the work of Galileo and Newton to that of Planck, Einstein, Rutherford, Bohr, Pauli, de Broglie, and Heisenberg. Descriptions of some of the great experiments in physics are included because they illustrate the method and spirit of scientific endeavor. From time to time the reader is reminded to step aside and, as would a philosopher, "think about man's thinking," to consider the roles of definitions, laws, and postulates, and the criteria for a successful theory, and to realize what sort of questions physicists try to answer.

The role of mathematics is also stressed. However, every effort has been made to keep the mathematical level low and still maintain a sound presentation. When derivations are simple and employ only algebraic substitution, these are given as examples of the rigorous methods of theoretical physics. As a balance, the experimental side is presented through the inclusion at the end of a chapter of one or more appropriate laboratory experiments. Theory and procedure are given for each, thus eliminating the need for separate laboratory instruction sheets. It is not intended that all of the experiments should be done, but that a choice of interesting ones be available.

In most nontechnical colleges and universities, students who do not major in science take a laboratory course, either a year of one science or a semester each of two sciences. If this course is to be their only glimpse in college of what science is like and what it is about, is it not wise to make this course one which does not require the memorization of soon-forgotten facts, but which does impart

v

an understanding of the methods and principles of physical science? Should not a knowledge of the concepts and basic postulates formulated by the world's greatest scientists of the past and present be of value to those who may become leaders of business, industry, government, the law, and other professions? This book is based upon affirmative answers to these questions. Physics is taken as the basic science and its laws are shown to form the foundations of chemistry, astronomy, and physical science in general.

Chapter 1 serves as an introduction. It is also intended to arouse the student's interest. In a fourteen-and-one-half-week term it should be possible to devote one week to each of the remaining thirteen chapters, with time left for tests and review. The chapters are pretty much independent of one another; each is built around a central theme. The steps leading to the discovery of our fundamental laws and the way in which the physicist's viewpoint of these laws has changed are regarded as dramatic and exciting episodes in the development of man's thinking. The basic laws are illustrated by giving carefully worked examples and by discussing the applications of these laws to topics of current interest such as skiing, sailing, climbing, the motion of satellites, atomic accelerators, radio waves, mass-energy equivalence, space travel, lasers, radioactive decay, nuclear fission and fusion.

It is indeed hoped that readers of this book will want to explore further the exciting world of physics and to follow its future development, and that they will find the study of physics interesting and enjoyable.

Hartford, Connecticut   F. W. C.
February, 1967

Contents

THE CONSERVATION OF CHARGE AND COULOMB'S LAW OF ELECTROSTATIC FORCE   CHAPTER 7

AMPERE'S LAW OF MAGNETIC FORCE   CHAPTER 8

FARADAY'S AND MAXWELL'S LAWS OF ELECTROMAGNETIC INDUCTION   CHAPTER 9

CHAPTER 10   THE RELATIVITY PRINCIPLE

CHAPTER 11   THE QUANTUM PRINCIPLE

CHAPTER 12   PAULI'S EXCLUSION PRINCIPLE

THE CONSERVATION OF MATTER PRINCIPLE   CHAPTER 13

NUCLEAR FORCES AND THE SEARCH FOR ADDITIONAL PRINCIPLES   CHAPTER 14

1|1 THE IMPORTANCE OF SCIENCE TODAY

Today almost everyone is talking about science. As we look around us at the many devices that have been recently invented with the avowed purpose of making our lives more comfortable and interesting, we realize that we are indebted to science for this new way of life, and so science seems to be wonderful and beneficial. Then when we consider the terrible destructive power of modern weapons, also developed through science, we feel that science may not always work for the good of mankind. However, whether the accomplishments of science are for good or evil, we all admit their importance. For this reason alone, educated

Understanding Science

people naturally will, and should, want to know more about science: its nature, its methods, and its philosophy.

In the Soviet Union the importance attached to science is evident in many ways. Every high-school student must study physics not for one but for three or four years. A Russian scientist ranks near the top of the Soviet hierarchy in both salary and social prestige. Scientific achievements vie with athletic prowess in capturing the admiration and enthusiasm of the Russian public.

In the western world our regard for science is shown in different ways. Fortunately we are beginning to realize the importance of more education in science. We can see that good scientists are in great demand and that the supply is inadequate. We may also notice that our advertising world recognizes the importance of science; it goes to great pains to explain "scientifically" why one kind of cigarette or pill is superior to those of other brand names, or why the inclusion of a special ingredient, say "X-19," makes a certain brand of toothpaste or face cream excel all rivals. We are also quick to acclaim the startling new advances in science, such as those into outer space.

1|2 THE NEED TO UNDERSTAND SCIENCE

While nearly everyone admits the importance of science, the need to study and understand it is not so universally granted. There is a common feeling that science is difficult and that studying it should be left to those who have an aptitude for it and will become (or are) scientists.

There is no doubt that we need to train more good scientists. To do this we must find the young people who have the necessary ability, and we must interest them in science to the extent that they will further their study of it. This means that we must show these people the true nature of science and the excitement, not of mystifying demonstrations, but of real discovery and achievement. This is one purpose of this book.

A second reason for understanding the nature of science is gradually being realized. We are training our college students with the hope that they will be qualified eventually to occupy positions of leadership and prestige. Whatever the type of their job, the chances are great that an understanding of science will be an important asset to them. Consider the example of our industrial leaders: in the interests of economy they must be familiar with the latest technological developments, and they must assign an important share of their company's expenditures to research. Consider also our political leaders: they are constantly being confronted with scientific problems, such as those involving water power and other sources of energy, radio and television communication, military weapons, radioactive fallout, etc. Scientific committees are set up in our Congress, and other committees call in scientists to give expert testimony. Can our lawmakers understand these professional scientists? Not unless the scientists can express themselves clearly and their questioners have learned something of the *language of science*.

Modern science touches on all fields of thought and endeavor: religion, the law, the fine arts, social studies, etc.

Clerical authorities generally look upon the laws of nature as God's laws for the physical world, and the fact that scientists find these laws to be comprehensive, yet simple and few in number, furnishes a strong argument for belief in the omniscience and wisdom of God.

In legal affairs we find, time and again, that scientific methods are employed, and that scientific evidence enters into the testimony of a case.

2

In order to show the relation between science and the fine arts, we shall later contrast the objective view of sound (the description of the frequency, intensity, etc., of its waves) with the subjective view of the musician, and we shall contrast the objective view of color with the artist's subjective view. We shall see that the two viewpoints complement one another, just as a description of a man's physical appearance plus a description of his character and behavior gives a more complete picture of the whole man than would either description alone. As another example of the relation between science and the arts, take the poet who, in his way, describes the same relationships in our physical world that the scientist sums up in a mathematical law.

Finally we come to the social sciences, which, by their very name, claim a close relationship to the natural sciences. This is because they all employ, to a lesser or greater extent, scientific methods and reasoning, and because there are many analogies between concepts in, say, economics and concepts in physics. There is a delight in discovering such analogies, and there is also the danger of setting up false ones. Here we again see the need for understanding what science is.

A third reason for studying science in any liberal curriculum is that it is a method of inquiry, a way of going about the discovery of what the world around us is like. The distinction between science and the humanities is not so much one of subject matter as one of methods. There are situations that can best be met through faith. Some discoveries and some wise decisions are the result of intuition and hunches. The artist is guided largely by feeling and emotion, which are important parts of our experience. The scientific method is another important way of adding to our knowledge and of solving our problems. It also involves imagination and intuition, but its most important ingredients are reasoning without prejudice, and experimental observation. A scientist will say: "Let's reason this out." Then, if he cannot convince us through arguments involving the laws of physics and, quite possibly, some impressive-looking equations, he will resort to the experimental method and obtain data, taken from observations of the effect in question, to prove his point.

We can all benefit from an understanding of the methods of science. Such an understanding will help us to judge values, to sift ideas and recognize half-truths and propaganda for what they are, and thus to form our individual opinions.

While we should not pretend that science will, or can, solve all of the world's problems, we must admit that the methods of science can be of great help to us; learning these methods should, therefore, be a part of a liberal education.

3

In early 1958 the U. S. Senate Committee on Labor and Public Welfare conducted hearings on science and education for national defense. In the course of the hearings, President L. A. DuBridge of the California Institute of Technology said:

Again we hear the cry: "Do not forget the liberal arts!" To that, of course, there is a simple reply: Science *is* one of the essential liberal arts. It ranks along with literature, art, music, as one of the finest and most elevating achievements of the mind of man. A liberal arts education does not deserve the name if it includes no science.

During the same hearings Dr. I. I. Rabi of Columbia University said:

I feel that in this scientific age every person who has been through school should have some feeling for science, and some feeling for the world about him, because it is science which is moving and changing the world and if he is entirely ignorant of this he is just being tossed about.

It is true not only for the general public, and I feel very strongly it should be true for people who in the future will be in government, and who in the future will be occupying your seats, gentlemen, in the Senate and the Congress and the executive departments of the Federal Government and of the State governments and city governments and municipalities.

They simply have to have some kind of capacity to understand what is happening in the world of science, to be somewhat familiar with what it is all about.

1|3 WHY STUDY PHYSICS?

We have already discussed the reasons for understanding *science*. Now let us consider why we should study *physics* in particular. The reasons will be enumerated.

a | In science, physics occupies the most central position.

Physics is the key science. It employs logic and mathematics on the one hand, and it leads to such practical developments as engineering and technology on the other. Physics is also used in forming the basic explanations in chemistry, geology, meteorology, astronomy, etc., and physical methods and instruments are used in biology, oceanography, psychology, medicine, etc.

b | Physics is concerned with the basic laws of nature.

Consider the physical (inanimate) world in which we live; it appears to operate according to certain fixed principles, such as Newton's law of gravitation and the conservation of matter. We learn about these prin-

4

ciples from infancy through personal experience. We also learn from the accumulated experience of others. This knowledge is vital. We cannot alter the way nature works, but we can put its laws to our use if we know what they are. Thus we learn not to jump out of a window, but utilize the stairs when we want to descend from an upper floor. We put gravity to our use when we use a water tank to supply water to a town, or when we run a clock with a falling weight. We get pleasure from gravitational forces, which the law of gravitation describes, when we ski down a graded slope. So it is with the other principles of our physical world. Is it not, then, a good thing to study and understand these principles more thoroughly? That is what a physics course enables us to do, for the laws of nature form the basis for all branches of physics. Physics investigates the world in which we live.

c | **Technological developments follow from the fundamental discoveries of physics.**

To understand, more than superficially, the principles of radio or rocketry, we must first learn the basic laws of physics. In a scientific career a certain amount of "know-how" may qualify one to be a technician, but the top jobs and the more interesting ones require a knowledge of fundamentals.

d | **The study of physics is excellent mental training.**

The study of physics teaches us new ideas, how to think and reason, how to use logic and its close relative, mathematics. It stimulates our imagination; for instance, it presents models to help us picture and understand the physical world.

e | **Physics teaches us how to observe.**

We shall discuss presently the experimental method in which observations are made as accurately as possible and in a controlled manner. We shall also discuss how experimental and theoretical physics fit together.

1|4 PHYSICS AND PHILOSOPHY

The deeper understanding of the physical world is closely related to philosophy. Many excellent books have been written on this subject, and references to some of them are given at the end of this chapter. In this book, philosophical implications will be discussed as we go along. This is a traditional approach, for before science covered as much knowledge and became as specialized as it is today, it was customary to teach it under the name of "natural philosophy." It is important to retain some of this approach.

Philosophy has been defined as "man's thinking about his thinking." By this we mean that in philosophy, man tries to gain a greater perspective of his accomplishments by stepping aside, as it were, and looking at them more from a distance. So in physics we shall introduce definitions, units, various kinds of laws and theories, and then we shall also ask ourselves what these things really mean. What is a law? What is a theory? Does physics answer the question of how the world behaves, or the question of why it behaves as it does? The answers to these questions are philosophical.

Logic is a branch of philosophy. It is concerned with methods of reasoning and hence enters into physical thought. We shall encounter most frequently the following types of logic.

1. *Inductive* reasoning, in which one proceeds from specific cases to a general principle. The laws of physics are most often arrived at in this way.

2. *Deductive* reasoning, in which one proceeds from the general principle to the specific case. Whenever laws are applied to a particular problem, this type of reasoning is employed.

3. Reasoning by *analogy*, in which successful reasoning in one branch of physics suggests a similar line of reasoning in another area. For example, many mechanical systems have electrical and acoustical analogues, with similar theories for each.

4. *Abductive* reasoning, in which one postulates the existence of a fact quite different from anything observed, yet a fact from which, according to known laws, something observed necessarily results. A good example is Maxwell's postulate and theory of electromagnetic waves (Chapter 9).

The *syllogism* is a familiar example of logical reasoning. Perhaps the most famous syllogism is the following, which dates back to the Greeks:

All men are mortal. (P)
I am a man. (S)
Therefore, I am mortal. (C)

We call P the primary premise or hypothesis, S the secondary premise, and C the conclusion. If P and S are true, then C must be true. It is instructive to examine this syllogism more closely, because it involves a line of reasoning analogous to that frequently encountered in physics.

The first requisite for logical reasoning is a clear definition of terms. In the syllogism above we must know what we mean by *man* (or *men*); the Greeks undoubtedly took this term to include the human inhabitants, but not the gods, of their world.

The next step is to establish the validity of our premises. The primary premise P appears to be fairly sound because it has been observed to hold true in billions of special cases, without an exception. Thus by inductive reasoning we establish the hypothesis P. Premise S follows from the definition of man.

Our syllogism concludes with a step in deductive reasoning. The hypothesis P is applied to a special case, which is "I." What is the conclusion? It is that I and every other living person are mortal. But is this conclusion absolutely certain? We cannot be 100% sure of the premise P because we cannot be absolutely sure of the future. For instance, how do we know that new medical discoveries will not someday prolong life indefinitely? We may say that this seems most unlikely and not worth considering, which is equivalent to saying that P is very nearly 100% certain. Then the conclusion C must also be nearly 100% certain, and we might as well treat it as a fact and conduct our lives accordingly. If this conclusion seems to you to be too obvious to justify its proof, it is because you have already gone through the above reasoning many times, consciously or unconsciously. The results of this sort of reasoning continually affect the way we plan our lives; for example, we reason that the sun will light the earth tomorrow, that spring will follow winter, etc. Just as some common predictions, such as those about weather and climate, are less certain than others, so we shall see that physical predictions also have varying degrees of certainty.

The logic employed in physics does not usually follow the strict form of the syllogism, although it could frequently be put in that form if so desired. The above example was given mainly to illustrate what is meant by logical reasoning and to show that its conclusions are not always equally reliable. In inductive reasoning one assumes that what is known to be true in a finite number of cases is true in general. Since one can never test the validity of a postulate in an infinite number of cases, even if one could think of all such cases, inductive reasoning involves extrapolation, with its attendant risk of uncertainty. Hence when one applies such a postulate to a new, untested situation, the conclusion arrived at deductively is also subject to some uncertainty. For illustration let us take the famous episode of Sir Isaac Newton and the apple.

Whether or not the story of the apple is fiction, we do know that Newton was the first person to relate the force pulling the moon toward the earth (and that pulling the earth toward the sun) to the force that makes objects fall toward the ground. Through inductive reasoning he arrived at the universal law of gravitation. This law states that between any two material bodies there exists a mutual attractive force. Newton actually observed a gravitational force to exist only in those cases where one of the bodies was large, namely the earth or the sun. More than 100 years later Cavendish devised a delicate method of detecting and measuring the gravitational attraction between two metal balls of ordinary size (i.e., with diameters of a few inches). Cavendish not only verified that the law of gravitation applied to such bodies, but he also determined the strength of the gravitational force for two given bodies at a known separation. One could sum up the reasoning of Cavendish in the following syllogism:

All material bodies obey the law of gravitation. (P)
Metal balls are material bodies. (S)
Therefore, the law of gravitation applies to such balls. (C)

The first premise is Newton's brilliant hypothesis. Since it was only an "educated guess," the conclusion reached from it could not be regarded by Cavendish as certain until he had tried his experiment; however, Newton's great reputation made Cavendish feel that the chances were good that the conclusion would be verified. It was!

It is interesting to note here that some of Newton's hypotheses and the conclusions drawn from them have not met the test of experiment. For example, Newton reasoned about light in a way that amounted to the following:

Moving particles that are attracted toward a medium which they are about to enter will be speeded up and so move faster in the new medium. (Inductive reasoning.) (P)

Light consists of particles that are attracted toward a denser medium. (Abductive reasoning. Newton made this "educated guess" to account for the observed bending of a ray of light when it enters a new medium.) (S)

Therefore, light travels faster in a denser medium. (Deductive reasoning.) (C)

Many years after Newton's lifetime, it became possible to measure the speed of light in different media and with sufficient accuracy to check the above conclusion C experimentally. It was found that light travels more *slowly* in glass and water than in air, so that C is false! This means that either P or S is false. Now P is based on Newton's law of motion, which states that a force produces acceleration, and this law is still accepted as one of the fundamental principles of nature; hence the faulty premise must be S. The present view is that while light is emitted and absorbed as though it consisted of particles (of energy), light is propagated as a wave motion, and is not attracted by glass and water surfaces. Newton offered postulate S only as an unverified guess; his guess was wrong!

An element of uncertainty enters into all scientific conclusions and for this reason scientists must learn to be humble and to avoid making extremely dogmatic statements. Consider as an extreme example the tossing of a coin. Will it land "heads" or "tails"? One cannot say because sufficient information, such as the exact manner of flipping the coin, is generally not given. If many truly random tosses are to be made, conclusions couched in statistical terms may be drawn, but even these predictions may not be borne out in a finite number of trials. In addition, mistakes in observing and counting are possible. So no physical prediction can be regarded as absolutely certain because man will never possess complete and entirely accurate information.

1|5 PHYSICS AND MATHEMATICS

While it is important to understand the physicist's point of view, or philosophy, it is equally necessary to be able to understand his language. Language is a tool for communicating ideas. In physics this is done through the use of our native language and mathematics. The latter has the advantage of being the *universal language;* it is used by scientists throughout the world.

Good science writing, whether or not it contains mathematics, should consist of clear and complete sentences. For example, suppose that we have defined a quantity x and then we define y as the square root of x and z as the reciprocal of y; we may then say:

"By definition (that of y in terms of x),

$$\sqrt{x} = y. \qquad \text{(a)}$$

"If we square each side of the equation, we get

$$x = y^2. \qquad \text{(b)}$$

"Also by definition (that of z in terms of y),

$$z = 1/y, \qquad \text{(c)}$$

so that (upon solving this equation for y),

$$y = 1/z. \qquad \text{(d)}$$

"If we substitute equation (d) in equation (b), we obtain the equation

$$x = (1/z)^2 = 1/z^2." \qquad \text{(e)}$$

9

The above quotation contains four complete sentences. The equality signs are read as "equals" and so play the role of verbs. The slash sign reads "divided by," a past participle, and y^2 reads "y squared." (The parenthetic statements are only for clarity and may be omitted.) We see that mathematics is in some respects a form of shorthand.

In physics, mathematical symbols are used to stand for ideas or concepts, which philosophers call mental *constructs*. Distance, time, and velocity are such quantities. The physicist defines a construct in such a way that it can be measured experimentally, either directly, or indirectly through the measurement of other related constructs. Philosophers call this measuring process *relating the construct to the plane of perception*. We shall explain later just how to measure distance, time, and velocity. For these quantities we will then substitute the symbols s, t, and v. These quantities will become part of our *vocabulary*.

The mathematical operations that we shall encounter are also defined and represented by symbols. We have already learned the meaning of such operational symbols as $=$, $/$, and $\sqrt{\ }$. Others will be explained when we have use for them. When used in our mathematical sentences, they are the *verbs* in our new vocabulary.

A mathematician sets up a branch of his subject in much the way one would invent a new game. He invents a set of self-consistent rules, and then he plays the game accordingly. Frequently, as in the case of Newton, a brilliant mathematician has also been a physicist, and then he has been careful to choose rules for his mathematical game which will make the resulting branch of mathematics useful in physics. But it has been interesting to note that even those branches of mathematics invented by "pure mathematicians," have also turned out to be useful to physicists. The rules or axioms introduced into arithmetic, algebra, geometry, calculus, vector analysis, etc., may be looked upon as additional *rules of grammar* in our mathematical language. Playing the game, or using the mathematics, must be done logically, just as a language must be used grammatically. It is also important that the game should be interesting.

Example. Since electronic computers employ units (transistors or tubes) which may be either "on" or "off," computations by such computers are based on a *binary system* of arithmetic in which only the digits 1 and 0 are employed. 1 indicates the condition "on" and 0 the condition "off." The rules of addition are

$$0 + 1 = 1, \quad 1 + 1 = 10, \quad 10 + 1 = 11,$$

$$11 + 1 = 100, \quad 100 + 1 = 101, \quad 101 + 1 = 110, \quad \text{etc.}$$

The rules of multiplication are the same as in ordinary arithmetic, that is,

$$0 \times 0 = 0, \quad 0 \times 1 = 0, \quad 1 \times 0 = 0, \quad 1 \times 1 = 1.$$

In the binary system 9 becomes 1001, since

$$9 = 1 \times 2^3 + 0 \times 2^2 + 0 \times 2^1 + 1 \times 2^0,$$

while 25 becomes 11001, since

$$25 = 1 \times 2^4 + 1 \times 2^3 + 0 \times 2^2 + 0 \times 2^1 + 1 \times 2^0.$$

Similarly, the product

$$25 \times 9 = 225$$

is expressed in binary notation as

11100001,

a step the reader should verify. Now as a check let us multiply 11001 by 1001, applying the rules for binary arithmetic. We get

```
    11001
     1001
    11001
11001
11100001.
```

So we see that we have a self-consistent set of rules.

Other examples of mathematical games are presented as puzzles at the end of this chapter. In these, the rules of the game may not always be stated explicitly but rather be left to common sense. In problem 7, for example, it is assumed that in a poker game one may win over others but not over himself, and that a theoretical physicist must be *good* at mathematics. Do *not* be discouraged if you can solve only two or three of these problems; they are not physics and you may not care for puzzles and games. However, when one does solve problems such as these he is sure to find fun and satisfaction in the accomplishment and see that logic and mathematics can be interesting and enjoyable. This is an important step toward success in studying physics.

In physics the "rules of the game" are not man-made, but they exist as the laws of nature, for us to discover and learn. This is really the purpose of physics. It is fortunate for us that nature's laws are also self-consistent and unchanging so that we may deduce conclusions from them in a logical way. Returning to the analogy between theoretical physics and a language, we should realize that the laws of physics are additional rules of grammar which should not be broken in our theoretical deductions. However, new discoveries have at times forced man to realize that his statements of nature's laws are not complete or perfectly correct. Hence as man reframes his statements of these laws our "rules of grammar" in physics become modified as well, just as they are in a living language.

Now try the problems and see whether you can discover some of the fun, as well as the spirit, of science.

PROBLEMS

1. a)   Prove that in a game of ordinary ticktacktoe it is always possible for either player to avoid defeat.

b)   How would you extend the game to three dimensions?

c)   How should the game be played in one dimension if it is always to end in a draw?

2.   Three men are blindfolded and told that either a red or a green hat will be placed on each of them. After this is done, the blindfolds are removed. Each man is asked to raise a hand if he sees a red hat and to leave the room as soon as he is sure of the color of *his* hat. All three men raise a hand, but for several minutes no man leaves. Finally one man, more astute than the rest, gets up and goes out. What color was his hat, and how did he reason it out?

3.   An explorer is in a region inhabited by people, each of whom either tells the truth all the time or lies all the time. The explorer comes to a fork in the road and wants to find out which road will lead him to the main village. He spies a native and, pointing to one of the roads, asks the native this question: "If I were to ask you if this is the road to your main village, would you answer Yes or No?" Explain whether or not the explorer will get the information he desires. Assume that the explorer does not know whether the native is a truth teller or a liar.

4.   The explorer in problem 3 meets two of the natives of the strange tribe of truth tellers and liars. "Are you a truth teller?" he asks the taller one. "Goom," the native replies. "He say 'Yes,'" explains the shorter native, who speaks English, "but him big liar." Find out whether each native spoke the truth or lied.

5. In the following, addition of two four-digit numbers yields a five-digit sum. Each letter represents throughout one and only one numeral. What are the numbers involved?

```
  SEND
  MORE
 MONEY
```

6. In the following, a four-digit number divided by a two-digit number yields a three-digit quotient. Each letter stands for one and only one numeral. Find the numbers involved.

```
AN | EASY | ONE
     AN
     UNS
     OES
      VBY
      VBY
```

7. Smith, Jones, and Robinson are the engineer, brakeman, and fireman on a train, but not necessarily in that order. On the same train are three passengers; we shall call them "Mr. Smith," "Mr. Jones," and "Mr. Robinson." One of the passengers lives in Chicago, one in Omaha, and one in Los Angeles. Further rules of the game are:

a) Mr. Robinson lives in Los Angeles,

b) Smith beats the fireman at poker,

c) the brakeman lives in Omaha,

d) Mr. Jones is no good at mathematics,

e) the passenger with the same name as the brakeman's lives in Chicago,

f) one passenger, a distinguished theoretical physicist, lives in the same town as the brakeman.

Who is the engineer?

8. At a ski resort in Vermont there are four cottages in a row to the left of the main lodge. Barn Cottage is at the extreme left.

a) Mr. and Mrs. White are in Barn Cottage.

b) Mr. and Mrs. Black like to go fishing.

c) Milk is drunk by the family in the main lodge.

13

d) Mr. Green drinks bourbon.

e) The Lodge is immediately to the right of Chicken House Cottage.

f) The family from Boston likes to go swimming.

g) The family from Greenwich is staying in Court House Cottage.

h) Scotch is drunk in the middle building.

i) The Brown family is staying in the first cottage beyond the Lodge.

j) The family from New York is staying next to the family of golfers.

k) The family from Greenwich is staying next to the tennis players.

l) The family from Hartford drinks iced tea.

m) The Blue family comes from Philadelphia.

n) The Browns are staying in a cottage next to Butterfly Cottage.

o) Each family stays in one building, comes from one place, has one favorite sport and one favorite beverage.

Answer this: Which family likes to hunt and which one drinks gin?

9. "I hear some youngsters playing in the backyard," said Jones, a student in mathematics. "Are they all yours?"

"Heavens, no," exclaimed Professor Smith, the eminent number theorist. "My children are playing with friends from three other families in the neighborhood, although our family happens to be the largest. The Browns have a smaller number of children, the Greens have a still smaller number, and the Blacks the smallest of all."

"How many children are there altogether?" asked Jones.

"Let me put it this way," said Smith. "There are fewer than 18 children, and the product of the numbers in the four families happens to be my house number which you saw when you arrived."

Jones took a notebook and pencil from his pocket and started scribbling. A moment later he looked up and said, "I need more information. Is there more than one child in the Black family?"

As soon as Smith replied, Jones smiled and correctly stated the number of children in each family.

Knowing the house number and whether the Blacks had more than one child, Jones found the problem trivial. It is a remarkable fact, however, that the number of children in each family can be determined solely on the basis of the information given above! What is the solution?

10. A person arranges the cards in one or more decks according to a preconceived rule. Can you find the rule in each of the following two cases

(S stands for spades, H for hearts, D for diamonds, and C for clubs):

a) 3H, 6H, 9D, JS, 8S, 7S, 5C, 4D, 9S, 6D, JH, QH, KC, 10H, JD, QS, 2D, 3C, 2H, 4H, 7H, 8H, 9C, 8C, 6C, . . . ?

b) 3H, 5S, 6C, 10C, 9C, 7D, 4H, 9H, 2C, 8C, JC, 6S, 9S, QH, 5H, 5D, 5C, 7S, 7H, 4D, 8D, QD, 3D, 3S, . . . ?

The solution of this sort of problem requires inductive reasoning. You must work toward the general rule according to which each given sequence of two consecutive cards is possible.

STATISTICAL FLUCTUATIONS IN THE BACKGROUND RADIATION **Experiment 1**
An Example of Uncertainty in Physical Measurement

Object: To observe and measure the fluctuations in the background radiation and to compare these measured fluctuations with those predicted by theory as most likely.

Theory: In some physical experiments one measures quantities that are believed to have a definite value, such as the speed of light in air, the charge on an electron, the melting point of zinc, or the value of π. For such quantities repeated measurements lead to what we call their "accepted values," values good within the limits imposed by the accuracy of the measuring equipment and the care taken by expert observers. If your measurement of π, say, differs from the accepted value, you should be able to justify your actual "error" by showing that it arose because your method was not sufficiently sensitive nor your instruments sufficiently accurate and that an even greater error was possible.

Suppose, however, that one tosses a coin 10 times and counts the number of times heads are observed. The observations can certainly be made accurately, but you may get 6 heads, another student 5, another 4, etc. Is 5 the "accepted" value? If you got 6, was your result in error? The new element in this sort of experiment is the presence of random fluctuations that can only be handled statistically. The theory of probability enables us to compute the chances of getting 10, 9, 8, . . . , 0 heads when a coin is tossed 10 times, but in 100 such trials even these computed chances may not be borne out experimentally. Evidently nothing is certain in this sort of experiment! There are fluctuations about the most probable value (5), and fluctuations in the fluctuations.

There are such random fluctuations to some extent in all physical measurements, but often only in the third or fourth digit of the value of a quantity. However, in the case of the background radiation, randomness

15

is even more complete than with well shuffled decks of cards. This background radiation includes (1) the secondary particles produced in our atmosphere and buildings by the cosmic radiation in space, (2) the radiation from radioactive contamination in the laboratory and radioactive minerals in the ground, and (3) radiation due to the fallout from atomic bombs released in the past several months. We shall measure the individual rays that pass through a Geiger counter (the operation of which your instructor may explain to you). Since one cosmic ray presumably travels through space quite independently and unaware of other such rays, we see that in a given time interval several such rays may happen to pass through our counter, while again few may do so in the same time interval. The counter may register 0, 1, 2, 3, ... counts in the given interval. It is interesting to observe the counts for many successive 10-second intervals (6-second intervals are better with a decimal minute timer) and to see how many times the count is 0, 1, 2, 3, ..., respectively. The results may then be compared with those predicted statistically.

Procedure

Step 1. Find the cosmic ray count after successive 10-second intervals for 20 or 30 minutes. The best procedure is for one man to watch the clock and call out "Read" at the end of each 10 seconds. Another man reads the glow tubes, and a third records data, etc.

Step 2. After the data has been taken, subtract each reading from the following one. Next prepare a chart with headings "0, 1, 2, 3, ..." Now if during the first 10 seconds you got 6 counts, put a mark under "6" on your chart. Continue for all the 10-second intervals in your data.

Step 3. Count the number of times (N) you got 0, 1, 2, 3, ... counts, and plot these numbers (N), versus the respective count numbers (n), that is, against 0, 1, 2, 3, ... Connect your points on the graph with a smooth curve. What count number n occurred most often?

Step 4. Compute the average number of counts per 10 seconds. How does this number compare with your most probable count number?

Step 5. If we let $\bar{n}$ equal the average number of counts per interval and A equal the total number of intervals during the 20 or 30 minutes of the run, then probability theory tells us that the most likely distribution is the one for which

$$N = \frac{A}{e^{\bar{n}}} \cdot \frac{(\bar{n})^n}{n!}. \tag{1}$$

Here $n!$ means $1 \cdot 2 \cdot 3 \cdot 4 \cdot 5 \cdots n$ and e is the base of natural logarithms

(2.718). Those mathematically inclined may use Eq. (1) to compute N for $n = 0, 1, 2, 3, \ldots$; they may then compare their experimental values of N with those values which statistical theory says are most likely to be observed. Does the difference seem reasonable? Under what circumstances would you expect better agreement?

References

BRIDGMAN, P. W.,   *The Nature of Physical Theory*, Dover, 1936.

CAMPBELL, N.,   *What is Science?* Dover, 1952.

CONANT, J. B.,   *On Understanding Science*, Yale University Press, 1947.

DEREK, J., AND
PRICE, S.,   *Science Since Babylon*, Yale University Press, 1961.

GARDNER, M.,   *The Scientific American Book of Mathematical Puzzles and Diversions*, Simon and Schuster, 1959.

—————,   *The Second Scientific American Book of Mathematical Puzzles and Diversions*, Simon and Schuster, 1961.

KASNER, E., AND
NEWMAN, J.,   *Mathematics and the Imagination*, Simon and Schuster, 1940.

MARGENAU, H.,   *The Nature of Physical Reality*, McGraw-Hill, 1950.

2|1 THE SCIENTIFIC METHOD

Modern science owes its progress to a combination of experimental and theoretical work. Each has been a stimulus to the other.

We can say that in the development of a science the starting point must be experimentation. We cannot speculate and theorize until we have observations to speculate about. These observations may be simply those of our everyday lives, from infancy on, or they may result from planned and controlled experiments. Gradually the facts observed become so numerous as to be confusing and hard to handle. Until we find some pattern or order in the accumulated data, we do not know in what direction to turn next. At this point one looks for a correlation between two or more

Experiment and Theory

quantities. These quantities must, of course, first be defined, giving us a working vocabulary. The observed relations between our concepts constitute the rules of grammer of our subject. As previously stated, a self-consistent set of rules may be arbitrarily chosen in a branch of mathematics, but in a subject such as physics we must take the rules as we find them from our observations; they are the *laws of nature*.

The Greeks observed the world around them and noted how nature behaved, yet they failed to discover any of what are now considered the fundamental laws of physics. Why was this so? The reason was that the Greeks just let events *happen* in the random and complicated way that events in nature always take place. They did not try to *control* the way things happened, because that meant doing something with their hands and manual labor was considered demeaning for educated Greeks of the upper class—it was "beneath" them. The Greeks did much thinking and arguing about the physical world, but they did very little real experimenting. The work of Archimedes, which led him to his famous principle about the loss of weight of a body in a fluid, was an exception. Aristotle, on the other hand, just looked about, noted that leaves fell faster than feathers and stones faster than leaves (Fig. 2–1), and then concluded that a heavier body always fell faster than a lighter one; had he tried to drop together two heavy objects, such as two stones of different weight, he would have been surprised to find that they fell at the same rate. Unfortunately Aristotle did not experiment.

FIG. 2–1
Stones fall faster than leaves.

The Romans excelled in engineering, but they were not interested in what we now call basic research. During Roman rule, Greek knowledge was cut off from the West.

In the 12th century Greek scientific writings reached the West via the Moslems. The spark of learning was kept alive by the Church, which under Thomas Aquinas merged Greek scientific philosophy and Christian theology. Scholars were taught to accept without question the statements of Aristotle. These statements became authoritative dogma. They were not tested. Since Aristotle and Ptolemy had said that all the heavenly bodies revolved around our earth, that hypothesis (for that is what it was) had to be accepted without any suggestion of an alternative view if one did not wish to find himself in trouble with the Church. In such an atmosphere science could not advance.

Galileo, who lived around 1600 A.D., is generally credited with introducing the experimental method that embodies the spirit of modern science, and physics as we know it today dates mostly from that time. Galileo not only experimented, but he performed *controlled experiments*. He also showed the importance of being able to brush aside trivial but complicating factors and getting to the heart of the problem at hand. For example, friction and heat losses can never be completely eliminated in actual experiments, but they can be progressively reduced in importance until one can, at least in thought, determine what would happen if they were absent entirely.

With the development of the experimental method it became possible to discover more relationships between quantities and this led to the development of theories to explain these relationships. Since the most successful theories are based on fundamental postulates about our physical world, these fundamental laws or principles have been emerging in the past three hundred years. With knowledge of these principles at hand, man has been able to develop their applications and so we have the many inventions of the modern world.

Perhaps it is better not to say that there is *a scientific method*, but rather that there are certain ingredients in modern scientific procedure which have accounted for its success. These include the controlled experiment and the thought experiment, unbiased judgments, awareness of surprising events that happen fortuitously, and the ability to correlate things learned in one branch of physics with facts and observations encountered in other areas of the subject. One of the prime objectives of this book is to bring out these points so that by understanding them the reader will better understand the nature of science itself.

2|2 FUNDAMENTAL QUANTITIES AND UNITS

First we must learn the language used by scientists. Scientists try to be very clear in their definitions of terms in order that such terms will always mean the same thing; only in this way can scientists communicate with one another, as well as with nonscientists, and only in this way can a body of scientific knowledge be built up and passed along from one generation to the next.

In a science such as physics, the choice of what concepts are to be defined and how they are to be defined is made by the people who are developing a new branch of the subject. While these pioneer workers are free to make arbitrary choices, they are naturally guided by the principle that their definitions must be useful. Most frequently scientists define quantities which are constant (1) for a given object, or (2) for a given material, or (3) for the entire observable world. Thus length is a quantity which has a fixed value for a particular swimming pool, density has a nearly constant value for water and another constant value for aluminum, while the elementary unit of charge is always the same. A quantity is usually worth defining if it will help us to *relate other quantities*, provided that this relationship is of some interest. Physicists do not introduce new definitions just for the fun of it, or to give students more to learn! For example, one could define a new quantity, let us call it a person's "spread," as the ratio of the person's weight to his age. However, this quantity would vary tremendously from person to person and for the same man or woman it would be continually changing; even its value at a given instant might not be too informative, because what may be a healthy "spread" for a two-year-old child would not be for a grown man.

We cannot define a quantity that relates other quantities unless the latter have already been defined. Where do we start? The answer is that we must commence with a few quantities whose meaning we understand *intuitively* and simply define how these *fundamental quantities* are to be measured. This means that units for our fundamental quantities must be chosen and specified. Then all other quantities and their units will be defined in terms of the fundamental ones.

Just which quantities we take as our fundamental ones is arbitrary, but experience has shown that the most useful scheme is to choose three in mechanics and, when needed, *temperature* and *electric charge* (or *electric current*). The latter will be defined later. For the three in mechanics we shall choose *length*, *mass*, and *time*.

a | Length

Our senses give us the concept of distance and extension in space. Such quantities were first measured crudely by using the spread of one's hand, the length of one's foot, etc. Then it was decided to adopt a standard of length and this was chosen to be one ten-millionth of (10^{-7} times) the distance between the equator and poles of the earth. The trouble with such a standard is that it is not easy to measure it accurately, and so as time went on and better methods were found, the length believed to correspond to the defined standard had to be changed. Finally, it was decided to make a rigid bar of noncorrosive platinum-iridium alloy and mark on it two fine lines separated by a distance approximately equal to 10^{-7} times the equator-to-pole distance; the distance between these two lines was then adopted as our standard unit of length, and it is called *one meter* (abbreviated to m). The meter is now legally defined in this way. The standard bar is kept in the *Bureau des Poids et Mesures* in Sèvres, France.

Standard meter bars have been prepared for other countries by direct comparison with the one in France. Secondary standards may be checked against the standard one, cheaper measuring sticks against secondary standards, and so on. In case something might happen to the standard

22

bar, its length has been related to the wavelength of a particular spectral line emitted by the atoms of a given chemical element. In the laboratory such spectral lines can be reproduced very accurately.

b | Mass

With our senses we recognize the existence of matter, and so we want to be able to measure the quantity of it in an object. To do this we use the mass of a body. While mass will be defined concisely in Chapter 3, where it will be indentified with a body's inertia to a change in its state of motion, suffice it to say here that the mass of a body is very nearly proportional to the number of heavy elementary particles (*neutrons* and *protons*) contained in the body. Thus in this sense the mass of a body is a measure of the quantity of matter in the body.

The present standard of mass, which is called *one kilogram* (abbreviated to kg) is taken to be that of a piece of platinum alloy also kept at Sèvres, France. Standards of mass for other countries, secondary standards, tertiary standards, etc., are prepared by comparison. The standard kilogram was chosen so as to have a mass nearly equal to that of a cube of water one-tenth meter on a side.

c | Time

Our senses make us aware of the passage of time. We observe certain rhythmic events, such as the beat of a pendulum. If we use this to mark time, we find that other physical events repeat themselves at fixed time intervals, and this makes us feel that a pendulum clock is a good device for indicating equal time intervals. Your heartbeat, on the other hand, is a poor timer because when we compare our heartbeat with periodic phenomena in the physical world (pendulums, vibrating springs, rotation of the earth, etc.) we find that either our heartbeat rate changes, or else all the above-mentioned physical phenomena suffer changes in period. We rule out the latter possibility as unlikely and recognize that our heartbeat probably varies. What physical phenomenon, then, shall we take to measure time? The present standard is based on the *mean solar day*, which is divided into 24 hours of 3600 seconds each. The *second* (abbreviated to sec) is the universal unit of time in physics. More recently it has been found that certain "atomic clocks" are more scientific timekeepers than our rotating earth, but for our purposes we may take the second to be 1/86,400 of the mean solar day.

The meter, kilogram, and second, together with the units derived from these, form the *mks system*. This is a metric system which has been almost universally adopted by physicists. Its units also find general use in most countries other than the English-speaking ones. While our common electrical units fit into this sytem, the British and American mechanical units, except for the second, do not. Generally we shall use the mks system, but in problems involving simply length, mass, and time (not force), we may use any appropriate unit for each of the three quantities. However, it is not acceptable to use two different units for the same quantity in a given calculation. The British-American units have been defined in terms of the mks ones, as follows:

0.3048 m = 1 foot (ft)

0.454 kg = the mass of, or quantity of matter in, an object weighing one pound (lb) at sea level.

Once these definitions have been memorized, one may easily work out others, such as:

1 inch (in.) = 2.54×10^{-2} m,

1 mile (mi) = 1.61×10^{3} m,

1 kilometer (km) = $\frac{5}{8}$ mi

1 kilogram = the mass of an object weighing 2.2 lb at sea level.

So if you ask for *un kilo de fromage* (one kilogram of cheese) in France you will get a quantity equivalent to what you get when you ask for 2.2 *pounds of cheese* in an American grocery store.

2|3 DERIVED QUANTITIES AND THEIR UNITS

In order to illustrate how we define quantities other than our basic ones, the definitions of some derived quantities that are useful in mechanics will be given. Other quantities will be defined as we come to them; it will be our rule to define each one in terms of previously defined quantities.

a | Area

The area (A) of a square is defined as the square of the length of a side. Other areas are defined and measured by counting the number of unit squares and fractions thereof contained in the given area.

b | Volume

The volume (V) of a cube is defined as the cube of the length of one side. Other volumes are defined and measured by counting how many unit cubes and fractions thereof are contained in them.

c | Speed

If a body travels a distance s in a time t, its *average speed* $(\bar{v})$ is defined as s/t, or

$$\bar{v} = \frac{s}{t}. \qquad\qquad \textbf{2-1}$$

The *instantaneous speed* (v) is defined as the limit of the ratio s/t when t is made very small. Speed will be regarded as the magnitude of *velocity;* the latter term also includes the direction of the motion.

d | Acceleration

If a body is moving in a fixed direction and its speed changes from v_1 at the time t_1 to v_2 at the time t_2, its *average acceleration* $(\bar{a})$ in the given direction is defined as

$$\bar{a} = \frac{v_2 - v_1}{t_2 - t_1}. \qquad\qquad \textbf{2-2}$$

When we consider constant accelerations in this book, we may take Eq. (2–2) as also giving us the definition of *instantaneous acceleration* (a). Let us write Δv for $(v_2 - v_1)$ and Δt for $(t_2 - t_1)$; then

$$a = \frac{\Delta v}{\Delta t}. \qquad\qquad \textbf{2-3}$$

The symbol "Δ" means "change in."

e | Period

The period (T) of an event that repeats itself at regular intervals of time is the time of one such interval.

f | Frequency

In periodic phenomena the frequency (f) is defined as the number of periods in unit time, or

$$f = \frac{1}{T}. \qquad\qquad \textbf{2-4}$$

Thus if the period is $\frac{1}{10}$ sec, the frequency $f = 10$ per second $= 10/\text{sec}$.

g | Density

The density (ρ) of a body is defined as the ratio of its mass to its volume, or

$$\rho = \frac{m}{V}. \qquad\qquad \textbf{2-5}$$

What are the mks units for derived quantities such as the above? The answer to this question is to let the definitions define the units. In doing so we adopt the following rule.

RULE: When physical quantities are multiplied or divided, or raised to the nth *power, their corresponding units are respectively multiplied, divided, or raised to the* nth *power, like algebraic quantities.*

As a result of this rule the units on each side of a physical equation must be essentially the same, which makes good sense. (Sometimes more than one name is given to the same unit or combination of units.)

Application of this rule to the definitions given above shows that the mks units for these quantities are as follows.

a) The *mks unit for area* is the square meter (m^2):

$$1 \, m^2 = (10^2 \, cm)^2 = 10^4 \, cm^2.$$

b) The *mks unit for volume* is the cubic meter (m^3):

$$1 \, m^3 = (10^2 \, cm)^3 = 10^6 \, cm^3 = 10^3 \, liters.$$

c) The *mks unit for speed or velocity* is the meter per second (m/sec).

d) The *mks unit for acceleration* is the meter per second per second (m/sec^2).

e) Period is a time, so its unit is the second.

f) Frequency is the reciprocal of time and its unit is the per second ($sec^{-1} = 1/sec$).

g) The *mks unit for density* is the kilogram per cubic meter (kg/m^3).

Summary

Note that each new quantity was defined in terms of our fundamental quantities and/or derived quantities previously defined. Observe also that when symbols for the various quantities are introduced, the definition of a derived quantity may be expressed as an equation, which may be combined with other equations in physics. Finally, remember that when we measure the magnitude of some quantity, say the length (l) of a table, our answer must consist of two parts, namely (1) the unit(s) employed, and (2) the number of such units contained in l. Thus our answer might be 2.54 m, which would be equivalent to 254 cm, 100 in., 8.33 ft, 2.78 yards (yd), etc. We see that the numerical part of our answer may have various values, such as 2.54 and 2.78, but for a given choice of units there is only one correct numerical answer.

Example. Express (a) 1.80×10^{12} cm/min and (b) 2.7 gm/cm^3 in mks units.

Solution

a) $1.80 \times 10^{12} \dfrac{\text{cm}}{\text{min}} = \dfrac{1.8 \times 10^{12} \times 10^{-2} \text{ m}}{60 \text{ sec}}$

$= 3 \times 10^8$ m/sec.

b) $2.7 \dfrac{\text{gm}}{\text{cm}^3} = \dfrac{2.7 \times 10^{-3} \text{ kg}}{(10^{-2} \text{ m})^3}$

$= 2.7 \times 10^3 \dfrac{\text{kg}}{\text{m}^3}.$

2|4 EXPERIMENTAL PROCEDURE

Having defined certain quantities, the physicist proceeds to look for relations between them, usually by performing controlled experiments. A controlled experiment is one in which the experimenter controls all the factors which may affect the result and then deliberately varies one such factor at a time. In this way one can find out how the result of the experiment (the effect) depends on each of the factors (the causes) involved. It is obvious that if one allows simultaneous variations in several relevant factors, then it will be difficult to sort out the individual contributions which each variation made toward changing the result of the experiment. Let us take an example.

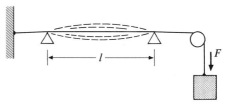

FIG. 2–2
A vibrating string.

Consider a stretched steel wire or string supported by two bridges (points of support) that are a distance l apart. Let the force pulling the string taut be F (Fig. 2–2). The string when plucked will vibrate with a particular frequency f. Now if one simultaneously loosens the steel string (decreases F) and shortens the distance l, the frequency f may either increase or decrease, depending on how much one loosens and how much one shortens the string. Just loosening a violin string lowers its pitch and just shortening the vibrating length raises the pitch. To sort out these two effects one should (1) keep l constant and vary F, and then (2) keep F constant and vary l, noting each time the variation in the frequency f. One will then discover that f is directly proportional to the square root of F and inversely proportional to l, that is,

$$f \propto \frac{\sqrt{F}}{l},$$

2-6

or

$$f = k\frac{\sqrt{F}}{l},$$

2-7

where k is independent of F and l. Suppose that F and l are each halved; will the frequency go up or down? (It will go up. Why?) Or suppose that F is increased to four times its original value and l is doubled, how will f be affected?

Experimental investigation of vibrating strings may be further extended by varying the diameter of the string. To do this properly one should compare the frequencies of two strings of a given material of equal length l which are under the same tension F, but which have different diameters. Note that on musical instruments, thick strings give notes of lower frequency than do thin strings. All else being the same, the frequency is inversely proportional to the diameter d, so that we may write

$$f = k'\frac{\sqrt{F}}{ld},$$

2-8

where k' is independent of F, l, and d.

The experimental investigator of vibrating strings would seek to find out whether the frequency depends on any other factors besides F, l, and d. He would finally discover that f also depends on the density ρ of the substance of which the wire is composed and that

$$f = \frac{k''}{ld}\sqrt{\frac{F}{\rho}},$$

2-9

where $k'' = 1/\sqrt{\pi}$, so that k'' is a universal constant for all vibrating strings. This would seem to indicate that consideration has now been given to all the important factors on which f depends. Further testing would show that variations in atmospheric pressure, temperature, elevation above sea level, color of string, etc. alone have negligible effects on the frequency. With this much information in hand, physicists would usually feel that a theory is called for to explain the dependence of f on F, l, d, and ρ and why k'' equals $1/\sqrt{\pi}$. How such a theory is constructed will be explained presently.

In everyday life we frequently lose sight of all the factors that must be controlled if we are to draw scientific information and valid conclusions from experiments. A person may alter (1) his diet, (2) the amount of sleep he gets, and (3) the amount of exercise he does, with the result that he feels better. To what should he credit his greater well-being? It would be unscientific and risky for this man to claim either that all the factors altered were equally important, or that one was particularly effective. Similarly, scientific claims should not be made by a tobacco company about the effect of altering the filtration in its cigarettes if at the same time the company alters by an unknown amount the strength of the tobacco used in the cigarettes. Only one factor at a time should be varied, and that by a measured amount.

The controlled experiment requires care and skill, but that is not all. The experimental physicist is confronted with the problem of sifting out the relevant factors. He must not only *control* but he must decide *what to vary*. One must acquire a method or *technique*, rather than experiment at random, and this requires intelligence and a knowledge of one's subject. Experiments 3 and 5 at the end of this chapter and the next are good examples of controlled experiments.

2|5 EMPIRICAL LAWS

We saw in the example of the vibrating string that by means of controlled experiments scientists have found a relationship between the frequency f and the tension F, length l, diameter d, and density ρ of a string, as stated by Eq. (2–9). Such a relationship, based purely on experimental evidence is called an *empirical relationship*, or an *empirical law*. Empirical laws must be distinguished from those postulates or hypotheses made by the theoretical physicist, although such postulates are also often called "laws."

Empirical laws are only as good as the experimental observations on which they are based; they stand subject to contradiction when one uses them to guess the result of an experiment that has not been tried before. As an example, consider the stretching of a coiled spring (Fig. 2–3). With a steel spring that is not stretched too far, one will find that the stretching force F is directly proportional to the resulting elongation s, or

$$F = ks, \qquad\qquad 2\text{--}10$$

where k is constant for a given spring but varies from one spring to another.

The statement of the above proportionality, or the statement that k is independent of s for a given spring, is often called *Hooke's law*. Now we know that rubber bands are quite stretchable and elastic, so we might be tempted to assume that Hooke's law also applies to the stretching of rubber bands, but Experiment 4 will show that this is not the case. Nor does Hooke's predict the breaking point for a stretched spring. Similarly the empirical law pertaining to the frequency of a stretched string does not apply when the stretching force goes to zero, or is so small that the string is not taut. So we see that an empirical law may have a limited range of applicability.

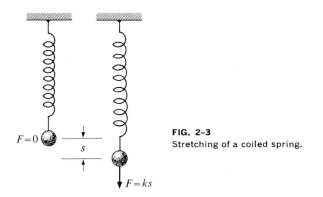

$F=0$ s

$F=ks$

FIG. 2–3
Stretching of a coiled spring.

The empirical law of Bode and Titus is another interesting example. It refers to the spacing of the planets in our solar system. Although the planetary orbits are elliptical, they are only slightly so and one may speak of the *mean radius* $\overline{R}$ of a planet's orbit, or the average distance of the planet from the sun. On the basis of astronomical data known around the turn of the last century Bode and Titus suspected the existence of some regularity in the values of $\overline{R}$ for the planets then known. Adopting the scale $\overline{R} = 1$ unit for Earth, the experimental values are: $\overline{R} = 0.39$ unit for Mercury, 0.72 unit for Venus, 1.5 units for Mars, 5.2 units for Jupiter, and 9.5 units for Saturn. The empirical law proposed by Bode and Titus was

$$\overline{R} = 0.4 + 0.3 \times 2^{n-2}. \qquad\qquad \textbf{2–11}$$

Here n represents the planet's *number*, counting out from the sun, that is, $n = 1$ for Mercury, $n = 2$ for Venus, $n = 3$ for Earth, $n = 4$ for Mars, while for the larger outer planets one must let $n = 6$ for Jupiter and $n = 7$ for Saturn. It was also stipulated that for $n = 1$ the second term in Eq. (2–11) was to be omitted; this rule and the omission of $n = 5$ are characteristic of the artificiality of some empirical laws. However, with the introduction of only two arbitrary constants (0.4 and 0.3), Bode and Titus got for *six* planets a remarkable fit between the experimental values for $\overline{R}$ and those computed from Eq. (2–11). Furthermore, this law made some successful and noteworthy predictions. For one thing it emphasized the gap between the inner and outer planets, since the planet for which $n = 5$ seemed to be missing. The search for the "missing" planet led to the discovery of the asteroids, small planetary bodies (the largest, Ceres, has a diameter of less than 500 miles) which circulate in orbits for which $\overline{R}$ is about 2.8 units. Note that in Eq. (2–11) $\overline{R} = 2.8$ when $n = 5$.

The Bode-Titus law also helped direct astronomers in their search for and discovery of the outermost planets, namely Uranus (for which $n = 8$ and the fit is good), Neptune, and Pluto (for which the fit is quite poor).

Many other examples of empirical laws will be encountered in the following chapters.

2|6 THE ACCURACY OF MEASUREMENTS

It is difficult to use relatively inaccurate data to find an empirical law and any relationship so discovered cannot be regarded as firmly established. Imagine, for instance, that one performs an experiment in order to determine whether a quantity y depends on another quantity x and, if so, what the dependence and constant(s) of proportionality are, and that one obtains the following data: $y = 0$ when $x = 0$, $y = 12$ when $x = 3.0$, $y = 21$ when $x = 6.0$, and $y = 36$ when $x = 8.0$. It is apparent that y increases with x, but how? Is y equal to $4x$, or $3.5x$, or $4.5x$, or is there no linear relationship? "No simple dependence" would be the answer if the values of x and y are exact to the figures given. However, if x is measured to only ± 1.0 unit and y to ± 2 units, then the relationship $y = 4x$ would be possible but not firmly established, since other proportionalities are not ruled out. In this experiment it would be much better to measure x to ± 0.1 unit and y to ± 0.5 unit, in which case the relationship $y = 4x$ would lead to readings such as $y = 0$ when $x = 0$, $y = 12.5$ when $x = 3.0$, $y = 24.0$ when $x = 6.0$, and $y = 31.5$ when $x = 8.0$, from which the relationship is reasonably apparent.

As another example suppose that we time a pendulum whose length is 64 cm and arc of swing is 25 cm and find that 25 oscillations take 40.2 sec. Next we might change bobs, while trying not to vary the length of the pendulum or the arc of swing, and we might then find that 25 oscillations take 39.8 sec. We would immediately wonder whether the difference between the two times was significant and implied a dependence of the period of oscillation on the bob used. Here our conclusion would be based on two considerations, namely, (1) how accurately we could measure the two times and (2) how closely we could keep the other factors (length of the pendulum and arc of swing) under control, and hence constant. If our timing was only good to half a second, then the difference between 40.2 sec and 39.8 sec would not be significant. If we could not change bobs without altering the length, the change in time might be due to a change in length, because further investigation would disclose a dependence of period on length ($T \propto \sqrt{l}$); actually a 2% change in length would account for the 0.4 sec or 1% change in time for the 25 oscillations.

It is evident that the experimental investigator must have a clear comprehension of the probable accuracy of his experiment and of the errors that may enter into it. As a complete study of errors is lengthy and involved, a simplification is necessary for our purposes. It is better to arrive easily at a fairly good understanding of the accuracy of an experiment than to undertake a detailed treatment of errors which is so complicated that it is never completed!

Inaccuracies in the results of an experiment may be due to (1) the method chosen, (2) the physical environment, (3) the observer, and (4) the equipment used. When one tries to estimate mathematically the possible error in his results, he finds that the important distinction is whether errors are (a) *systematic* or (b) *random;* therefore, we shall adopt this classification. Systematic errors include all inaccuracies that tend to be more in one direction than in the opposite one. Random errors are those that may be treated statistically, positive and negative errors being equally likely and large errors less probable than small ones. Taking repeated readings of a quantity and averaging reduces the random errors, but not the systematic ones, hence care and pains should be taken to eliminate or correct for systematic errors. Let us now consider the latter.

First there are *mistakes*, which may be due to carelessness, misreading a scale, lack of understanding of the equipment, faulty arithmetic, miscounting, etc. Mistakes *must* be eliminated.

Systematic errors due to the *method* can, once they are discovered, be reduced to insignificance or corrected for. For example, when one measures temperature with a glass bulb thermometer, one should calculate the effect of the expansion of the glass and correct for it if necessary. If such a thermometer is calibrated by placing it in ice water and then in boiling water, one must be sure that the ice bath is not a little above the freezing point and the hot bath not a little below the boiling point, otherwise the interval between these two calibration points will be reduced. If the circumference of a cylinder is measured by wrapping a tape around it, a slight error will arise due to the finite thickness of the tape; this may be estimated and corrected for, or one may use thinner tape, or one may be able to use larger cylinders for which the error is less percentagewise.

As far as the *environment* is concerned the important thing is that it be properly controlled.

The idiosyncrasies of a particular *observer* may be checked by having more than one person take readings.

Instrumental errors of a systematic nature may be most important. Examples are: a warped or shrunken meter stick, a clock that always runs slow (or fast), a damaged electric meter that always reads low (or high), an instrument whose zero point is off, etc. It is imperative to eliminate such sources of error by checking all questionable pieces of equipment against one or more reliable instruments. Faulty apparatus should be discarded, repaired, or calibrated and supplied with a table of corrections.

Let us now assume that systematic errors have been rendered insignificant. One should next ascertain whether a reading is *reproducible*. If it is, then one should allow for a possible error of one part in the smallest unit read or of one division of the instrument, whichever is larger. Remember that a manufacturer usually puts divisions on his scales to match the accuracy of his instruments. From this it follows that one should carefully choose instruments whose ranges are appropriate, that is, instruments such that the readings encompass a good fraction of the divisions on the full scale. Thus a meter stick calibrated in millimeters is an accurate instrument for measuring a distance of 65 cm (650 mm), but not for measuring a length of 1.4 cm (14 mm), because 1 mm in 650 mm is about $1\frac{1}{2}$ parts per thousand (0.15%), while 1 mm in 14 mm is about 7 parts per hundred (7%).

When readings are, for one reason or another, not reproducible, it is necessary to find out how much they vary before one can estimate their reliability. Suppose that we make six observations of some quantity and get numerical readings that are bunched closely together, e.g, 20.2, 20.1, 20.0, 20.3, 19.9, and 20.1. Would we not feel considerable confidence in our data and in taking the mean value of 20.1 as our answer? On the other hand, suppose that our six numerical readings were 22.2, 18.0, 24.2, 15.3, 23.0, and 17.9; how much confidence would we now feel in our readings or in their mean value of 20.1? Very little! For one thing, we can see that if we had omitted any one observation, then our mean value would have been quite different, while if we had taken a seventh reading, that also

33

would probably have altered our mean value. Evidently what must be considered here are the fluctuations in the readings. The significance of these fluctuations is most easily evaluated by calculating the *average* or *mean deviation* $\bar{d}$. The deviation d for any one reading is here defined as the difference between the reading and the mean, with the smaller quantity being subtracted from the larger so as to make d come out positive. To obtain $\bar{d}$ one must average the deviations for all the n readings, so that

$$\bar{d} = \frac{d_1 + d_2 + d_3 + \cdots + d_n}{n}. \qquad\qquad \text{2–12}$$

The theory of probability tells us that, assuming no systematic errors, for a set of six readings ($n = 6$), 95% of the time the actual error in the mean will not exceed $\bar{d}$; henceforth we shall usually assume systematic errors eliminated and regard $\bar{d}$ as our possible error due to random fluctuations, provided that at least six readings are taken.

Example 1. Take the two sets of data given above. Compute the possible error in each case.

Solution. It is best to tabulate as in Table 2–1. For the first set of data $\bar{d} = 0.1$, and the random possible error is 0.5% of the mean. For the second set of data $\bar{d} = 3.0$, and the possible error is about 15% of the mean. (To check these percentages note that 1% of 20.1 (the mean) is $0.01 \times 20.1 = 0.2$.)

TABLE 2–1

Reading	Deviation	Reading	Deviation
20.2	0.1	22.2	2.1
20.1	0.0	18.0	2.1
20.0	0.1	24.2	4.1
20.3	0.2	15.3	4.8
19.9	0.2	23.0	2.9
20.1	0.0	17.9	2.2
Sum 120.6	0.6	120.6	18.2
Mean 20.1	0.1	20.1	3.03

In the first case it would seem very unlikely that the true value of the quantity being measured would actually lie above 20.2 or below 20.0, while for the other set 20.1 ± 3.1 would seem to provide a liberal range of uncertainty.

Frequently the final result of an experiment is obtained by taking sets of observations of several different quantities and then adding, subtracting, multiplying, dividing, etc., the various mean values. For example, we might use Eq. (2–9) to compute the frequency f. Knowing the percent errors in our measured value of F, l, and d, respectively, how would we find the percent error in our computed value of f?

When we combine quantities, the important rules to remember are:

1. *When adding or subtracting quantities, add the numerical errors.*

2. *When multiplying or dividing quantities, add the percent errors.*

3. *When raising a quantity to the nth power, multiply the percent error in the quantity by n.*

4. *When extracting the nth root of a quantity, divide the percent error in the quantity by n.*

Example 2. Given $x = 200 \pm 2$, $y = 160 \pm 2$, $z = 2.00 \pm 0.05$. Find the possible error in $(x - y)/z^2$.

Solution. We have $x - y = 40 \pm 4$. The percent errors are 10% in $x - y$, 2.5% in z, 5% in z^2. Therefore the percent error in $(x - y)/z^2$ is $10 + 5 = 15\%$. We may write

$$(x - y)/z^2 = 10.0 \pm 1.5.$$

Example 3. Refer to Eq. (2–9). Suppose that our measurements give $F = (100 \pm 2)$N, $l = (50.1 \pm 0.2)$ cm, $d = (50 \pm 2) \times 10^{-2}$ mm, and $\rho = (8.1 \pm 0.1)$ gm/cm^3. Compute the possible percent error in f.

Solution. The percent errors are 0.4% in l, 4% in d, 2% in F and 1.2% in ρ. One may take π to as many correct figures as desired so that its error may be neglected. Then the percent error in f is

$$0.4 + 4.0 + \tfrac{1}{2}(2.0 + 1.2) = 6.0\%.$$

2|7 THE ESTABLISHMENT OF EMPIRICAL LAWS

Now that we have seen how the reliability of a measurement may be evaluated, we return to the discussion of how experimental relationships (empirical laws) are established. This may involve the discovery that a certain quantity does not change in value when the value of another quantity is varied, or it may involve the discovery of a functional relationship between two quantities.

a | Establishment of a relationship of independence between two quantities

Suppose that we want to determine whether the time of 25 swings of a pendulum is independent of the arc of swing, or whether the time for a stone to fall from rest through 20 ft is independent of the size of the stone, or whether the ratio of the circumference to the diameter of a cylinder is independent of the size of the cylinder. How would we proceed? In the case of the pendulum, we could take six readings of the time with one arc and six readings with the same pendulum but a larger arc. For each set of conditions we would find the mean time and the mean deviation. Then we would compute the possible error in each case. Only if the mean times for the two arcs differed by *more* than the sum of the possible errors should we definitely conclude that altering the arc of swing did affect the period of the pendulum.

Mean times of (40.2 ± 0.2) sec and (39.8 ± 0.2) sec should be regarded as possibly the same, within the accuracy of the experiment, since the two times might both be 40.0 sec. On the other hand (40.4 ± 0.2) sec and (39.8 ± 0.2) sec should not be regarded as the same.

The equality of two measurements made under different conditions may frequently be established by a *null method*, that is, a method in which *no effect* indicates equality. If little or no effect is expected, one may replace the measuring device with a much more sensitive detector. For example, two pendulums of equal length, but with different bobs, could be set in motion together and an electrical device could be used to detect whether or not the two bobs continued to swing in synchronism.

FIG. 2-4
Galileo dropping stones from the Leaning Tower of Pisa.

Galileo used the null method in his famous experiment on falling objects. He dropped (legend says from the Leaning Tower of Pisa) a small and a large stone together and visually compared their rates of falling

(Fig. 2–4). Crude though Galileo's technique may now seem, it was truly scientific, and it led him to the discovery of some important properties of the physical world. Such experiments help us to see more clearly *how* nature behaves.

b | Establishment of a relationship of proportionality between two quantities

Next, suppose that we want to prove that the distance s through which a stone falls in the time t is proportional to some power, call it the nth, of t, so that

$$s = kt^n, \hspace{4cm} \textbf{2-13}$$

where k is the constant of proportionality. We would also like to find experimentally the values of n and k. We might start by choosing a carefully measured distance s_1 and taking several readings of the time of fall; the mean value of the times will be called t_1, and we can compute its possible error. This one pair of readings of s and t will not tell us the values of either n or k, so we must choose another value of s, say s_2, and measure the corresponding value t_2 of t. Let us suppose that we do this for at least four different distances of fall. Our data may now be presented most clearly in tabular form, as in Table 2–2, which contains a typical set of numerical data. Should we average the values of s in Table 2–2 and find their mean and mean deviation? This would be useless because we know that the values of s are not supposed to be the same, while in the case of t we strongly suspect that the times increase with s, and so are not constant either. We should average values of a quantity only when we have reason to believe that the values vary only because of random fluctuations. At this point we should look carefully at our data and do a little thinking.

TABLE 2–2

s, ft	t, sec
$s_1 = 16$	$t_1 = 1.04 \pm 0.05$
$s_2 = 32$	$t_2 = 1.40 \pm 0.05$
$s_3 = 48$	$t_3 = 1.70 \pm 0.05$
$s_4 = 64$	$t_4 = 2.01 \pm 0.05$

Remember that more often than not the relationships found in nature are simple ones. In the case of the freefall experiment, we should observe immediately that (1) t increases with s, and (2) t does not increase as rapidly as s does. This means that if an equation such as Eq. (2–13) exists, n must be positive and greater than unity $(n > 1)$. The simplest possibility, and one which the data strongly indicates, is that $n = 2$, or that s is proportional to t^2. Let us write

$$s = kt^2. \hspace{4cm} \textbf{2-14}$$

Next we should use our data to verify the proportionality between s and t^2 and to determine the value of k. This is best done graphically. If possible one should always seek a straight line graph because it is the easiest to draw; its slope will quickly give us the proportionality constant, and any departure from the expected proportionality is most apparent. So in our example we should plot t^2 against s. However, since no measurements are perfectly accurate, we cannot expect the points on our graph to lie *exactly* on one line. If the points do not fall on a line, how can we tell whether this is because of errors or because we have assumed the wrong functional relationship? The answer to this question lies in consideration of the possible errors in our data.

Table 2–3 lists the values of t^2 that correspond to the four values of s, according to the data of Table 2–2. The possible errors in each value of t^2 may be found by getting the percent error in t and doubling it to get the percent error in t^2. Or, taking t_2, for example, we can simply compute either $(1.40)^2 - (1.35)^2 = 0.14$, or $(1.45)^2 - (1.40)^2 = 0.14$, and see that $t_2 = 1.96 \pm 0.14 \text{ sec}^2$. Note that we may assume that when $s = 0$, $t = t^2 = 0$, as a stone requires no time to fall zero distance.

TABLE 2-3

s, ft	t^2, sec^2
0	0
16	1.08 ± 0.10
32	1.96 ± 0.14
48	2.89 ± 0.17
64	4.04 ± 0.20

After all our data have been plotted, as in Fig. 2–5, we should draw through each point a vertical line whose length above and below the plotted point represents the possible error in the measurement. The length of the line may be made clearer by ending it with bars, as in our figure. Now we can take a straightedge and easily see whether or not a line may be drawn that will pass through some part of each error line associated with our experimental points. If not, the assumed proportionality is not verified. If we find that many lines may be drawn, the range in slope of these lines will indicate the possible error in the constant of proportionality (see Fig. 2–5). In this way we would, in the free-fall experiment, arrive at the possible error Δk in k. Our best value for k would be that given by the line that best fitted our experimental points. In our particular graph this line passes through the origin and the point $s = 64$ ft, $t^2 = 4.00$ sec^2 and yields the value $k = \Delta s/\Delta t^2 = 64$ ft/4 sec$^2 = 16$ ft/sec^2.

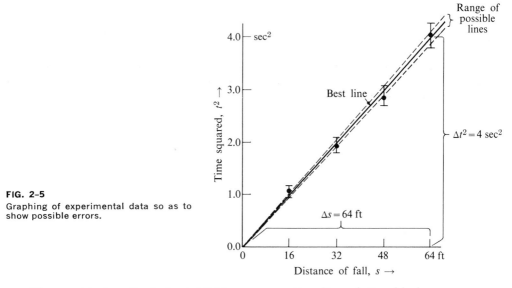

FIG. 2–5
Graphing of experimental data so as to show possible errors.

The graphical method of establishing a proportionality relationship is to be preferred over a purely numerical one. A graph quickly gives a comprehensive picture of the results. In the example just discussed, it is better to determine k graphically than by averaging values of s/t^2 computed from Table 2–3.

2|8 THE CONSTRUCTION OF A THEORY

We all theorize from time to time. For example, suppose that when you get in your car you find that it will not start. You immediately ask yourself what the trouble might be. Perhaps it is something that you can fix yourself, or, if not, you would at least like to be able to give a garage man, when you call him, some idea of what he should bring. So you make a guess as to the cause of your trouble; let us say that you suppose it has something to do with your battery. In science a guess of this sort is called an *hypothesis* or *postulate*.

The next step in the development of a theory is to reason from the hypothesis, to deduce from it certain consequences. The first such consequence must be what has been already observed experimentally. In the example about the car your guess about the battery will certainly explain why your car will not start. But of course there are many other possible reasons for this and you want to find the right one. Therefore you now *test* your hypothesis by considering some of its other consequences that may be checked experimentally. If your car trouble is in the battery, then what other symptoms should you look for? One would be that the horn and the lights should not work either; if testing shows that this is so, you will begin to feel more certain that your trouble really is connected with the battery and not with the ignition switch or the self-starter. Your next step might be to take a look at the battery. Perhaps you would find that the battery cable had been eaten away by acid, or that the cable was loose and not making a firm electrical connection. If the trouble is not in the cable, you might make the additional hypothesis that the battery is "run down." From this new guess you may deduce further consequences which may be tested experimentally, given the proper tools or equipment. One test would be to check the density (or specific gravity) of the solution in the battery, using for this purpose an instrument called a "hydrometer." Another test would be to use a voltmeter to find out whether or not the battery is supplying the proper voltage. If you do not have the necessary instruments or the experimental "know-how" to use them, you must call for expert help. But when a mechanic comes he will simply continue the method outlined above until he feels reasonably sure that he knows what your trouble is. If he is a good mechanic, he will make several tests of his hypotheses before he subjects you to the cost of expensive new parts and labor. Probably we have all experienced instances in which experts have guessed wrong. Physicists too have made some poor guesses in the past, and their theories have been superseded later on by better theories.

From the last section, we see that a theory involves making one or more hypotheses and then deducing the consequences. These consequences must include the facts that we already know. Thus the first aim of a theory is to *explain*, in terms with which we are familiar, the cause and effect relationship between our hypotheses and these facts. The *more* facts it correlates, the better the theory. We look upon theories as aids in our understanding of the way in which our world behaves. For example, in the eighteenth and nineteenth centuries it was popular to propose a *mechanistic* explanation for many nonmechanical phenomena. The internal energy associated with the temperature of a body was attributed to the random motion of molecules, sound was explained in terms of a to-and-fro motion of particles in the transmitting medium and Newton showed that even light *might* (he made it clear that this was only a guess) be pictured as a stream of particles that are attracted toward a denser medium, such as glass. These theories appealed to people because everyone was familiar with the motion of balls and other ordinary objects, and so they enabled everyone to visualize, as it were, what happens in nature when water is heated, when sound is transmitted, and when light passes from air into glass. Such theories represent a description of an invisible or otherwise not easily recognizable experience in terms of better known phenomena.

FIG. 2–6
The earth is attracted toward the sun by the sun's gravitational field.

Sun

Earth

The formation of a theory frequently involves the mental construction of an associated *model*. The sun's gravitational field may be pictured in terms of lines (they are called lines of force) radiating out in all directions away from the sun (Fig. 2–6). Of course such lines do not really exist. Another famous example of a physicist's model is the planetary model of the atom, which is associated with the Bohr theory of the hydrogen spectrum. In his theory Bohr adopted a model of the atom which was first proposed by Rutherford. According to this model, an atom consists of a positive core called the nucleus, surrounded by circulating negative charges called electrons (Fig. 2–7). The nucleus is supposed to contain nearly all of the mass of the atom. An electron is attracted to the nucleus of its atom by an electrostatic force which varies inversely as the square of the distance of separation, but the electron does not fall into the nucleus because it is supposed to be circling around it at sufficient speed to maintain a fixed orbit, just as does an earth satellite in the earth's gravitational field of attraction. Note that this whole model bears a very close analogy to our solar system, although the scale is vastly different and the attractive force is of different origin in the two cases. The nucleus is the atom's sun; the electrons are its planets or satellites. Since we are all rather familiar with drawings and descriptions of our solar system, we can immediately form a mental picture of Rutherford's atom. However, no one

41

has even *seen* an atom with its nucleus and circling electrons, and such a system may not really exist. In fact, other atomic models have been proposed, and currently Rutherford's model is not taken too seriously.

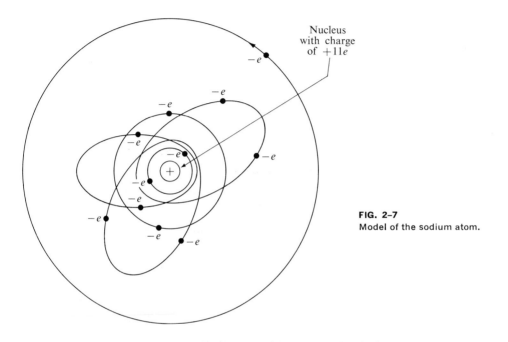

FIG. 2–7
Model of the sodium atom.

While the explanation of physical phenomena in terms with which we are more familiar is a worthwhile objective for a theory, it is not the most important goal from the physicist's point of view. To a physicist a theory is not successful unless it is *fruitful*, that is, the theory should enable one to deduce certain consequences which were not previously known but which, upon experimental testing, are found to be actual properties of our physical world. In this respect physical theories are far more than a process of diagnosing and explaining the cause of some effect; the main purpose of constructing a physical theory is to use it as a means of making new discoveries.

This is also true of models. The planetary model of the atom suggested many things, such as the possibility of removing an electron from a normal neutral atom. What should we expect to have left if a negatively charged electron is taken away from a system in which the positive and negative charges originally balanced? Should not the resulting system be positively charged? It is found that atoms seem to become what are called *positive ions* in just this way. Atoms that gain extra electrons become *negative ions*, and, since unlike charges attract (Chapter 7), a positive ion and a negative ion may come together to form a molecule of a very stable compound, as in the case of common salt (NaCl), in a crystal of which the sodium atoms have lost electrons to the chlorine atoms. Here the model leads us on into chemistry.

The first step after choosing the hypotheses and possible model for a theory is to express the hypotheses in the form of mathematical equations. Since some of the hypotheses in physics cannot be fully expressed except in terms of higher mathematics, a more advanced study of physics necessitates learning the theory of the calculus, differential equations, vector and tensor analysis, etc. In fact, Newton *invented* the calculus because he found that he could not express his law of motion properly except as a differential equation! However, we shall see that the fundamental laws of physics may be expressed in an understandable, if not in their most general, form without the calculus.

The next step in developing the consequences of a theory is to combine the equations expressing hypotheses with others of a similar nature and with ones expressing definitions of derived quantities. With the aid of mathematical manipulation, certain variables are eliminated and the derived relations obtained. Here, facility with mathematics is a great help.

A poorly constructed theory, one based on bad "guesses," may explain the known facts all right, but it will not prove fruitful. This was the case with the geocentric theory of Ptolemy and Aristotle which postulated that the sun and planets revolve about the earth. This theory should not be labeled a "wrong" description of the apparent motions of the heavenly bodies. However, we now prefer the heliocentric theory, which supposes that the planets go around the sun, because it is much simpler, because it may be explained in terms of principles (gravitation and Newtonian mechanics) of wide scope, and because it has enabled people to make such correct predictions as the existence of the outermost planets and how to place an astronaut in a chosen orbit.

A less successful theory may make predictions that are not borne out experimentally, or new facts may later be discovered which do not fit into the theory. Such theories do not meet the criterion of being able to *grow gracefully*. In such cases physicists do one of two things, namely, (1) they try to correct the old theory by modifying or adding to its postulates, or (2) they start anew by forming a fresh and completely different theory. The modified, or the new, theory will have additional consequences, which in turn may be investigated experimentally until eventually the newer theory may also meet with failure. Thus the process goes on and

43

science progresses. We may indicate this progression schematically as follows:

Experiment → Theory 1 → Experiment → Theory 2 → Experiment → · · ·

Note the interplay between experimental and theoretical work that is necessary if science is to advance. Although some physicists have marked mathematical ability and prefer theoretical work, while others can do things better with their hands and like experimental investigation, it is important that one who desires to be a good physicist should be somewhat familiar with both kinds of work. The theoretical worker must be able to visualize tests of his theories that may be carried out practically, and the experimental investigator will enjoy his work much more if he understands the theory that he is trying to prove or disprove.

2|9 FUNDAMENTAL AND RESTRICTED LAWS

The great laws of physics are those that express principles or relations which are independent of the specific properties of certain materials or objects. These laws will therefore be called our *fundamental laws;* they must be distinguished from those *restricted laws* which apply only to certain materials and only under a limited range of conditions. Fundamental laws are our starting points in the various branches of physics. They are often arrived at by the inductive method, in which one seeks that general principle which will explain the greatest number of specific facts. In such cases, fundamental laws are not wild guesses. On the other hand, the postulates of the quantum theory and of relativity were bold, inspired guesses differing radically from previous beliefs. It is said that when Planck first made his quantum postulate he could hardly believe it himself, although from it he was able to derive for the first time an important empirical (experimentally observed) law whose explanation had been attempted unsuccessfully by others.

Restricted laws are usually first discovered experimentally, so that they are initially empirical laws. Physicists generally believe that such laws follow from the fundamental ones provided that one has complete information about the structure of the material or medium involved. When such information is not directly available, one must make further guesses in the form of *postulates about the structure of matter.* For example, in kinetic theory one assumes a gas to be composed of hard elastic spheres in random motion. Such hypotheses are not as universal in scope as are our fundamental laws, but, like all postulates, they must explain known

44

facts and they should lead us to new discoveries. In the process of establishing a theoretical basis for an empirical law, other laws are often deduced as well. These *derived laws* are really theorems that come out of the theory. They may be universal in scope (e.g., the conservation of momentum, $E = mc^2$, etc.) and so on a par with, but related to, our other fundamental laws, or they may be restricted to certain media or situations (e.g., the formula for the velocity of a wave in a string).

The various "laws" (principles, hypotheses, postulates, theorems) of physics may be classified as follows:

1. *Fundamental laws or principles: hypotheses or postulates of a general nature (e.g., the law of gravitation). These form the core of each of the following chapters.*

2. *Restricted laws: empirical relationships of limited applicability (e.g., Hooke's law of elasticity).*

3. *Postulates about the structure of matter: educated guesses (e.g., the postulates of kinetic theory, the theory of solids, atomic structure, etc.).*

4. *Derived laws: fundamental or restricted relationships discovered mathematically from (1), (2), and (3).*

While the constants associated with restricted laws have values that are fixed only for a specific object or material under a limited range of conditions, the constants associated with the fundamental laws are *universal constants*. The latter include G, the gravitational constant; k_e, the proportionality constant in Coulomb's law for the electric force between static charges; c, the speed of light *in vacuo; h*, Planck's quantum constant; k, Boltzmann's energy constant; and e, the smallest unit of electrical charge. The measured values of these constants and others are listed in Appendix 1 for ready reference.

PROBLEMS

1. Find in mks units
   a) the area of a circle whose radius is 2 cm, and
   b) the volume of a sphere of 20 cm radius.
   c) Convert 1 mi/hr into m/sec.

2. Given that by definition *force* equals mass times acceleration and *work* equals force times distance. Find the mks units for force and for work.

3. The gravitational constant G is defined by the equation $F = Gm_1m_2/r^2$, where F is the gravitational force between the masses m_1 and m_2, and r is the separation of the masses.

a) Find the mks units for G.

b) If $G = 6.67 \times 10^{-11}$ mks units, what is its numerical value in terms of grams, centimeters, and seconds (cgs units).

4. A pendulum swings through a small arc, and 25 oscillations are timed six times. The six readings are 40.6, 39.4, 40.2, 39.6, 40.6, and 40.2 sec.

a) Find the mean deviation and the percent possible error due to random fluctuations.

b) If the timer is known to run 1% fast, what would you take as the most probable value for the time of 25 oscillations?

5. The pendulum in problem 4 is made to swing through a larger arc, and the times for 25 oscillations are found to be 41.4, 42.2, 41.2, 43.2, 40.8, and 41.4 sec. Show whether or not the data in problems 4 and 5 indicate a definite dependence of the period of a pendulum on the length of the arc of swing.

6. A certain meter has 200 divisions and readings are good to one division. What would be the percent instrumental error in computing a deflection covering

a) full scale,

b) half scale, and

c) a fifth of the scale?

7. Refer to Eq. (2–7). Suppose that you have found by measurement that

$$f = (320 \pm 1) \text{ vibrations/sec}, \qquad F = (40.0 \pm 0.4) \text{ newtons (N)}$$

and

$$l = (0.400 \pm 0.002) \text{ m},$$

find the value of k and the percent error in its determination.

8. By measuring the circumference c and diameter d for each of six cylinders, a student obtains the following values for c/d: 3.148, 3.163, 3.154, 3.147, 3.160, and 3.146. Find the actual percent error in the mean and the percent mean deviation. Give reasons why the actual error may, as here, exceed the mean deviation.

9. a) An electron of charge e will experience a force eE due to an electric field E and a force evB when moving with speed v across a magnetic field B. What mathematical relation must hold true when both an electric and magnetic field are present, and the force due to one cancels that due to the other?

b) After an electron has been accelerated through a potential difference of V volts (V), its speed v is given by the relation $\frac{1}{2}mv^2 = eV$, where m is the electron's mass. Combine this equation with the one derived in (a) and obtain an expression for e/m in terms of E, B, and V, quantities which are easy to measure experimentally.

10. According to the law of friction, the force of friction between a block of wood and the table on which the block rests is doubled when the weight of the block is doubled.

a) Is the law of friction empirical?

b) Is it a fundamental or a restricted law? Explain.

c) What constant appears in this law? Is it a universal constant? (The reader is referred to one of the standard comprehensive texts.)

DETERMINATION OF π **Experiment 2**
Possible Error versus Actual Error

Object: To use the measurement of π to learn about the meaning and determination of error in experimentation.

Theory: To determine π experimentally one need only measure the diameter d and circumference c of a cylinder. Then since $c = \pi d$, we have

$$\pi = \frac{c}{d}. \tag{1}$$

Since there is an experimental limit to how accurately one can measure c and d, the calculated value of π will be correct only up to a certain figure. Experience shows that it is good practice to repeat a measurement several times and then average the values found. This average or *mean value* is more reliable than any single value. Furthermore, the *mean deviation* of the single values from the mean value is a measure of the *possible error* in the averaged value. Large deviations suggest an unreliable result, small deviations convey confidence in the answer. Therefore we will make ten calculations of π, using a different cylinder each time, and then calculate the mean deviation. Suppose, for example, that we get the following ten

47

values of π; then the mean and mean deviation would be found as shown:

π	Deviation
3.122	0.029
3.164	0.013
3.184	0.033
3.106	0.045
3.122	0.029
3.134	0.017
3.202	0.051
3.156	0.005
3.172	0.021
3.148	0.003
Sum = 31.510	0.246 = Sum
Mean = 3.151	0.025 = Mean deviation

Random error: It is customary to express errors in percent.

$$\% \text{ possible random error} = \frac{\text{Mean deviation}}{\text{Mean value}} \times 100\%$$

$$= \frac{0.025}{3.15} \times 100\%$$

$$= 0.8\% \text{ possible random error}$$

Actual error: In research one may not know the correct answer, but only that it should not differ from the mean measured value by more than the possible error. In this, as in some of the other lab experiments, the correct value is known or given and the *actual error* may be computed. If it exceeds the possible random error, a systematic error is indicated.

$$\text{Measured value of } \pi = 3.1510$$
$$\text{Correct value of } \pi = 3.1416$$
$$\text{Actual error} = 0.0094$$
$$\text{Percent actual error} = \frac{0.0094}{3.1416} \times 100\%$$
$$= 0.3\%,$$

which is completely covered by the possible random error.

Procedure

Step 1. Measure c and d for six or more different cylinders. Compute the mean value of c/d. Find the percent possible random error and percent actual error for *your* determination of π. Should the latter exceed the former, look into and discuss possible *systematic* errors.

Step 2. *Alternative method of measuring π:* A quite different way of measuring π involves throwing sticks on a ruled surface and counting what fraction crosses a line. (The lines must be ruled one stick length apart.) Then, within the limits of statistical predictions,

$$\pi = \frac{2}{\text{Fraction crossing a line}}. \tag{2}$$

Toss 100 toothpicks or matches on a suitably ruled surface. Compute π from (2). What actual error did this method yield? How does the accuracy of this second method compare with that of the first method? Repeat, using 400 or more sticks.

THE PERIOD OF A PENDULUM
The Experimental Method

Experiment 3

Object: To illustrate the experimental method by finding how the period of a pendulum depends on various factors.

Theory: When an object on the end of a string swings down, it falls due to gravity, after which it swings up, falls back, and returns almost to the starting point, ready to repeat the cycle. Obviously, the *period T*, or time of a complete cycle (over and back), must depend on g, the acceleration due to gravity of a *freely* falling body. But before we can find out how T depends on g we must investigate whether T depends on any other factors. What might these factors be? They are (1) the mass of the bob, (2) the length of the arc or swing, and (3) the length of the string. Following the true experimental method, we vary only one such factor at a time so that we can see what the dependence on each factor is.

Procedure

Step 1. Using a string of 64 cm length (measured from the support to the *center* of the bob) find the time of 25 swings through a small arc (20 or 25 cm in all). Repeat four more times and determine the mean value of T and the possible error.

Step 2. Varying the arc. Repeat Step 1, but with a larger arc of 50 cm or more. Within your possible errors, is a dependence of T on the length of arc established?

Step 3. Varying the bob. Repeat Step 1 with a bob of different mass, but with the same length of string and the same arc. Within your possible errors, is a dependence of T on the mass of the bob established?

Step 4. Varying the length of the string. Repeat Step 3 (new bob, small arc) but for a string length of

a) 100 cm,

b) 36 cm.

How does T depend on the length l? Plot whichever will give a straight line: T against l, T against l^2, or T^2 against l, and find the constant of proportionality (which Newtonian theory says should be $4\pi^2/g$).

Experiment 4   HOOKE'S LAW
A Law of Limited Range

Object: To investigate the relationship between stretching force and the resulting stretch, in order to determine if there is any "law" relating these quantities.

Theory: We may always define a new quantity as the ratio of two quantities already defined. So let us define the "elastic constant" k of a spring as the ratio of the force F to the stretch s resulting from F. Then

$$k = F/s. \tag{1}$$

Now this definition does not make k a constant or assign to it any other properties. If we observe experimentally that k has the property of being constant for a given spring over a range of applied forces, then we will have established an empirical law for this spring. Such a law was discovered by Hooke for springs or straight wires of steel, brass, etc., and it is known as Hooke's law.

We shall see that Hooke's law has a limited range of applicability. It does not hold for many materials, even ones that are readily stretched, and for steel the law holds only up to a certain limit, called the "elastic limit." Our problem is to investigate the relationship between s and F for (a) a steel spring, and (b) a rubber band.

Procedure

Step 1. Fasten the steel spring to the support and hook onto the lower end of the spring the hanger for slotted weights. A meter stick is mounted behind the hanger. Measure the length of the spring and record the height of a certain point on the hanger.

Place a 100-gm-weight on the hanger and record the new height of the reference point on the hanger. Continue this process until eight 100-gm-weights have been added. *Note:* When adding a weight release it slowly and don't drop it onto the hanger. Plot the stretch s against the weight causing it. Did the spring obey Hooke's law throughout your test? When it did obey Hooke's law, what was the constant value of k in gm-weight/cm?

Step 2. Repeat Step 1 for the rubber band, only continue carefully adding weights until you break the band. If after adding a weight you observe a creeping elongation, wait a minute or so before reading the hanger position. If you do not have enough 100-gm-weights, carefully remove five and add a 500-gm-weight in their place.

Plot s against F for the rubber band. Note whether any portion of the graph is a straight line; if so, for this portion determine k from the slope, that is, redefine k as

$$k = \Delta F / \Delta s, \tag{2}$$

where Δs means a small change in s and ΔF a small change in F. Does Hooke's law apply at all to a rubber band?

Which stretched more easily, the steel spring or the rubber band? Which stretched the greater fraction of its original length?

Reference

SHAMOS, M. H., *Great Experiments in Physics*, Holt-Dryden, 1959.

3|1 GALILEO'S FAMOUS EXPERIMENTS ON MOTION

The medieval scholastics who followed in the path of Aristotle argued about the causes of motion, postulating that heavy "earthy" objects, such as stones, sought their most "natural" state, namely one of rest in the lowest possible position. This hypothesis was not wrong, for it explained many facts, but it led nowhere.

Galileo pointed out that it was better to study motion itself before trying to discuss the causes of motion. In so doing he developed the modern experimental method.

Mention was made in Chapter 2 of Galileo's work on falling bodies. While he was too keen an observer not to notice that a light object did not fall *quite* as fast as a heavier one,

Newton's Laws of Motion

he also observed that the variation in the time of fall was much less than the variation in the size or weight of the object; for dense, heavy objects there was little difference in rate of fall. Galileo realized that the fact that the rates of fall were nearly the same was more significant than the fact that they differed slightly. He sensed that in an idealized situation where the effect of the air could be considered negligible, the rates of fall would be *exactly* the same. Although those who followed Galileo developed vacuum pumps and with them came much closer to creating an actual airless condition, Galileo did not have to wait for this to happen before drawing his conclusions. By combining *thought experiments* with fact he saved much time. This method has been useful ever since, but only to those who can supply the necessary thought!

In his inclined plane experiments Galileo observed that hard spheres rolled straight down an inclined plane in such a manner that, for a given inclination, the gain in speed Δv during a fixed time interval Δt seemed to be constant (Fig. 3–1). A similar situation in modern life would be that of a car starting off in such a manner that after one second the speedometer reads 5 mph, after another second 10 mph, after a third second 15 mph, etc. The ratio $\Delta v/\Delta t$, or the gain in speed per second, is constant in such cases and its value measures how fast the object in question is speeding up. Galileo realized that $\Delta v/\Delta t$ was an interesting and important quantity and he therefore proposed to give it a name by calling it the *acceleration a*. He was the first to introduce this scientific definition of acceleration, a concept that had previously been only vaguely associated with motion at varying speeds.

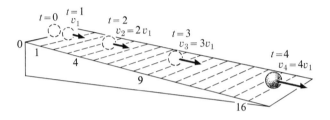

FIG. 3–1
Constant acceleration of a sphere rolling down an inclined plane.

Acceleration need not be constant, as it was in the examples above. A car's speedometer might read, in successive seconds, 35 mph, 40 mph, 43 mph, 45 mph, 45 mph, etc., indicating a decreasing acceleration as the car approaches a steady cruising speed. In such cases $\Delta v/\Delta t$ represents the *average acceleration* during the time Δt. Since, however, Galileo's experiments and ones, such as our Experiment 5, on freely falling objects (see Fig. 3–2) involve constant acceleration, we shall limit ourselves from here on to this important case.

Since by definition $a = \Delta v/\Delta t$, then for linear motion

$$\Delta v = a\,\Delta t \qquad\qquad \textbf{3–1}$$

represents the change in velocity of a body, during the time Δt, due to a constant acceleration a. If such a body starts from rest, then in t sec it will acquire a velocity given by Eq. (3–1). The Δ symbol means "change in," and it may be dropped for convenience if the body starts from rest when we start counting time and attains the speed v at the time t, so that $\Delta v = v$ and $\Delta t = t$; we may then write

$$v = at, \qquad\qquad \textbf{3–2}$$

where v is the velocity gained from rest in a time t under constant acceleration a. If the initial velocity is zero and the final is at and the change is at a uniform rate, then the *average velocity* $\bar{v}$ is the arithmetic mean, or

$$\bar{v} = \frac{0 + at}{2} = \tfrac{1}{2}at. \qquad \text{3-3}$$

The definition of $\bar{v}$ in terms of distance and time is

$$\bar{v} = \frac{s}{t}, \qquad \text{3-4}$$

where s is the distance traversed in the time t.* Let us equate the two expressions for $\bar{v}$; the result is

$$\frac{s}{t} = \tfrac{1}{2}at, \qquad s = \tfrac{1}{2}at^2. \qquad \text{3-5}$$

It was in this way that Galileo realized that when a body is accelerated from rest, at constant acceleration, s is proportional to t^2, or (symbolically) $s \propto t^2$ (see Fig. 3-1). He also showed that if $s \propto t^2$, then the acceleration must be constant, or

$$\Delta v \propto \Delta t.$$

Galileo rolled balls up and down planes at various inclinations. It was easy for him to see that the acceleration of a downward-rolling ball and the retardation of an upward-moving one decreased in magnitude as he made the slope of the plane smaller. He then raised the question:

What would the motion be if the plane were *very smooth* and *exactly horizontal?*

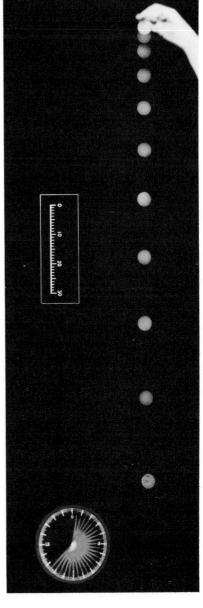

FIG. 3-2
A freely falling golf ball illuminated and photographed by light flashes equally spaced in time.

---

* Note that Eq. (3-4) is a definition and always applies, whereas Eq. (3-3) comes from the application of this definition to the special case of constant acceleration and hence Eq. (3-3) holds only for this case.

He answered this question by saying that the acceleration would be zero and that an object rolling or sliding on such a plane would move with constant velocity as far as the plane extended.

Of course Galileo could not eliminate friction entirely or be sure that a certain plane was exactly horizontal, but he could picture such a plane in his mind and conjecture about the behavior of a body on it. Here we see another example of the *thought experiment,* in which one mentally sets aside extraneous factors and concentrates on the point at issue, which in this case is how a body would move under the action of no force. Galileo was aware of the existence of friction and its retarding effect on motion along a horizontal plane, but he could see that as friction was reduced, the moving body kept on going for a longer and longer time. He realized that it took a force, such as friction, to stop the body, that is, change its state of motion. He thus arrived at the concept of *inertia,* of the tendency of a body to keep going when no force acts on it.

3|2 THE CONCEPT OF FORCE

Now let us consider what causes motion. Before the time of Galileo, scientists regarded force as a push or pull that was needed to keep a body moving. It was obvious that a steady force had to be applied to a plow, a boat, or a cart on level ground, just to maintain steady motion. But how can a force produce only steady, unaccelerated motion in the case of the cart, while in the case of a falling body, or a ball rolling down a plane, the force (pull) of gravity produces constant acceleration? The resolution of this paradox is to introduce the concept of *friction,* or to be more exact, *moving friction,* as a force opposing motion. We must then say that when a cart or sled (Fig. 3–3) is being pulled along level ground there are *two* horizontal forces acting in opposite directions, namely the forward pull of what we call the *applied force* and the backward drag of friction.

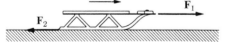

FIG. 3–3
The net force acting on the sled is the pull F_1 minus the frictional drag F_2. It is $F_1 - F_2$ that determines the change in speed of the sled.

What is the effect of these two forces together? We observe that the effect of friction is opposite to that of the applied force. Let us therefore define the *net force* acting on the cart as the difference between the two opposing forces and let us postulate that it is this net force that we must consider to be the effective force for *changing* the motion of the cart.

We see from the above that the direction of a force is important. Thus force must be regarded as a quantity having both *magnitude* and *direction*. Such a quantity is called a *vector*. Other quantities that are vectors are velocity, acceleration,* momentum, electric and magnetic fields. Quantities that have only a magnitude, such as length, mass, temperature, and energy, are called *scalars*. When we wish to emphasize the directional property of a vector quantity we shall use *bold-face type* for its symbol, for example, **F** for force. The symbol F will refer to the magnitude of a force only.

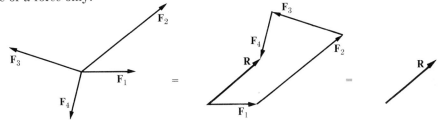

FIG. 3-4
Addition of several forces. The four forces together are equivalent to the single force **R**.

Whereas scalar quantities of a given kind are just added or subtracted numerically, experience shows that forces, as well as other vector quantities, must be added by placing, in any order, the directed lines proportional to each force end to end, as shown in Fig. 3–4; the single force **R** will have the same effect as all the individual forces, $\mathbf{F}_1$, $\mathbf{F}_2$, $\mathbf{F}_3$, and $\mathbf{F}_4$, acting together at a given point. Of course forces in the *same direction* are added numerically, while when two forces are in the *opposite direction* the numerical difference represents the effective force. For further understanding of the addition of forces, and vectors in general, see Experiment 6 at the end of this chapter. (See also Fig. 3–5).

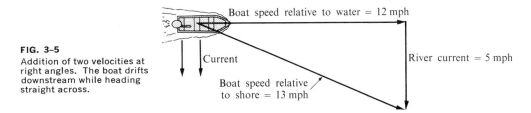

FIG. 3-5
Addition of two velocities at right angles. The boat drifts downstream while heading straight across.

---

* From now on acceleration will be regarded as the vector $\mathbf{a} = \Delta\mathbf{v}/\Delta t$, where $\Delta\mathbf{v}$ is that vector which must be added to the velocity at a given moment to obtain the velocity after the time interval Δt.

When a body rolls down a relatively smooth inclined plane, the acceleration is due to the pull of gravity. Since this pull is found to be vertically down toward the center of the earth, the part of the pull effective down the plane and hence the net force down the plane must be the projection, or what is termed the *component*, of the pull down the plane. This component decreases as the slope of the plane is made less (Fig. 3–6), but the acceleration also decreases with decreasing slope. Therefore, when constructing the concept of force, it would seem reasonable to associate net force with acceleration. The greater the acceleration of a given body the greater must be the net force acting on that body in the direction of the acceleration.

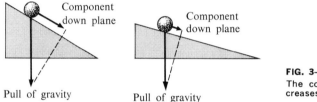

FIG. 3–6
The component of gravity down a plane decreases with decreasing inclination of the plane.

Is the net force acting on a body directly proportional to the resulting acceleration? We could define force so as to make this true, but there would then arise the question of whether this definition would be compatible with our intuitive conception of force in terms of push and pull.

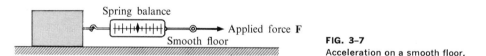

FIG. 3–7
Acceleration on a smooth floor.

Push and pull are usually measured with elastic springs that return to their original length when released. If we accelerated a body along a smooth horizontal surface and, as we did so, measured the force with a spring balance (Fig. 3–7) and the acceleration by measuring s and t in Eq. (3–5), we would find that for any given body the force as measured by the spring is proportional to the resulting acceleration. We may thus conclude that for a given body

$$\mathbf{F} \propto \mathbf{a}. \qquad\qquad 3\text{–}6$$

We may then think of force in terms of the acceleration it will impart to a body. We could measure force by measuring this acceleration; however, we would find that the ratio of F to a depends on the test body, so that we would have to adopt a standard test body. This is just what is done when we take as the *standard kilogram* a certain piece of platinum kept in Sèvres, France, with replicas in the bureaus of standards of other countries. For this standard kilogram we take the constant of proportionality in Eq. (3–6) to be unity (Fig. 3–8).

FIG. 3–8
Definition of the unit of force.

This defines our *mks unit of force*, which is called the *newton* (N). If a force **F** imparts an acceleration of x m/sec^2 to a standard kilogram, then **F** has a magnitude of x newtons, or x N. Thus

$1 \text{ N} = 1 \text{ kg-m/sec}^2$.

But what is the constant of proportionality in Eq. (3–6) when the test body is not a standard kilogram? The answer to this question involves Newton's concept of *mass*.

3|3 THE CONCEPT OF MASS

The ratio of the net applied force F to the resulting acceleration a was found by Newton to be a property or characteristic of a given body. This constant ratio is called the *mass* (strictly speaking, the *inertial mass*) of the body. For the standard kilogram we call the mass 1 kg. If a force **F** is needed to give a standard kilogram the acceleration **a**, while a second body requires a force m times as great to experience the same acceleration, then we say that the mass of the second body is m kg. We thus have

$$F \propto m \qquad\qquad\qquad \textbf{3-7}$$

for a given acceleration. If we combine Eq. (3–6) and Eq. (3–7) we get

$$\mathbf{F} \propto m\mathbf{a} \qquad\qquad\qquad \textbf{3-8}$$

for any object and any acceleration. Since we have chosen our unit of force so as to make F numerically equal to a when $m = 1$ kg, the constant of proportionality in Eq. (3–8) must be unity. Hence we have

$$\mathbf{F} = m\mathbf{a} \qquad\qquad\qquad \textbf{3-9}$$

for any object subjected to any acceleration.

Note that if the same force **F** is applied to two objects, one of mass m_1 and the other of mass m_2, then

$$m_1 a_1 = m_1 a_2,$$

$$\frac{a_1}{a_2} = \frac{m_2}{m_1}, \qquad\qquad \textbf{3-10}$$

or the greater the mass the less the acceleration. This is an *inverse* proportionality. It is for this reason that mass should be associated with *inertia*. The bigger m, the smaller a, and the harder it is to start the body in motion, or to slow it down once it is moving. Mass measures the reluctance of a body to change its state of motion or rest. If a lead ball and a wooden ball of the same size are thrown with equal force, the lead ball gains less speed than the other because it has the greater mass. If a heavy station wagon is to speed up and slow down as quickly as a light sports car, the engine and brakes of the station wagon must be capable of exerting greater forces than do the engine and brakes of the sports car. The sports car has less inertia and this is one reason why it is fun to drive it.

Is mass a measure of the quantity of matter in a body? The answer to this question is "very nearly so." For a long time the two concepts were regarded as synonymous. In fact, in everyday life a pound of butter or a kilogram of cheese represents to a customer in a grocery store a definite quantity of an item of food. When a substance changes its state, as when ice melts, the mass does not measurably change. Compressing a gas without letting any enter or leave the vessel does not alter the mass of the gas. In both of these cases the number of atoms or molecules remains constant. For a given chemical substance, mass is proportional to the number of atoms or molecules of the substance. The proportionality constant depends on the atomic or molecular weight of the substance. Since over 99.9% of the mass of an atom is the mass of its nucleus, and since nuclei are composed of neutrons and protons of almost equal mass, the mass of *any* object is very nearly proportional to the total number of *nucleons*, i.e., neutrons and protons, in the object. The constant is approximately 1.67×10^{-27} kg per nucleon for matter at rest or moving with ordinary speeds.

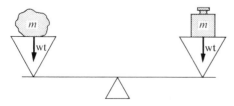

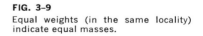

FIG. 3-9
Equal weights (in the same locality) indicate equal masses.

Weight is defined as the force of gravitational attraction acting on an object due to the celestial body on or near which the object is located. Weight depends on the location of a body, but at a fixed location is proportional to the mass of the body (see Section 4–6). For practical purposes the two pans of a balance are near enough together to be "in the same location" so that when the weights on the two sides balance we may take the masses on each side to be equal (Fig. 3–9). We thus use weighing as an easy method of *comparing* an unknown mass with a known mass. However, one should remember that weight is a *force*, not a mass.

3|4 NEWTON'S SECOND LAW OF MOTION. FUNDAMENTAL LAW I

Equation (3–9), $\mathbf{F} = m\mathbf{a}$, was founded on two proportionalities. It was stated as an experimental fact that for *any* given object the net force acting and the resulting acceleration are proportional. Mass was defined so as to make the force required to impart the same acceleration to different bodies proportional to the respective masses of the bodies. It is also an experimental fact that if this definition is used to determine the mass m of an object in terms of that of the standard kilogram, then the same value for m will be obtained *whatever* the acceleration used. In fact, the mass of an object is independent of time, of the position or acceleration of the object, and of the velocity of the object provided its speed does not approach that of light (see Chapter 10). This is true of *any* object whatsoever.

From the emphasis in the last paragraph on the words "any" and "whatever," it should be evident that the equation $\mathbf{F} = m\mathbf{a}$ not only summarizes a wealth of experimental information but also expresses something that is universally true. We may take any object, determine its mass by measuring its acceleration under a known force, mark the value of this mass on the object, and assume this to be the object's mass in future experiments. The realization of this universal principle dates from the work of Sir Isaac Newton (1642–1727).

Actually, Newton stated his principle in terms of *momentum* rather than acceleration. The momentum $\boldsymbol{\mu}$ of a body is defined as its mass times its velocity, or

$$\boldsymbol{\mu} = m\mathbf{v}. \hspace{3cm} \textbf{3–11}$$

Suppose that in the time Δt the velocity changes from $\mathbf{v}_1$ to $\mathbf{v}_2$; then the momentum changes from $m\mathbf{v}_1$ to $m\mathbf{v}_2$. Call this change in momentum

$\Delta\mu$. Now by definition $\mathbf{a} = (\mathbf{v}_2 - \mathbf{v}_1)/\Delta t$, so that Eq. (3–9) may be written as

I. $\quad \mathbf{F} = m\mathbf{a} = \dfrac{m\mathbf{v}_2 - m\mathbf{v}_1}{\Delta t} = \dfrac{\Delta\mu}{\Delta t}.$ \hfill 3–12

Following Descartes, who regarded mass times velocity (that is, momentum), as a measure of the "quantity of motion," Newton expressed his law in terms of this product, postulating that

for any object the net force equals the time rate of change in the momentum imparted to the body by the net force,

as stated mathematically by Eq. (3–12). This is often called *Newton's second law of motion;* it is our *Fundamental Law I* and will be referred to as "Newton's law of motion."

Example. What average net force will accelerate a 1000-kg rocket from rest up to a speed of 100 m/sec in 5 sec?

Solution. In Eq. (3–12) $\mathbf{v}_1 = 0$, so that

$$F = \frac{mv_2}{\Delta t} = \frac{1000 \text{ kg} \times 100 \text{ m/sec}}{5 \text{ sec}} = 2 \times 10^5 \text{ N}.$$

Since the acceleration $\mathbf{a} = 20 \text{ m/sec}^2$, the same answer is obtained by using $\mathbf{F} = m\mathbf{a}$.

Equations (3–9) and (3–12) are equivalent provided one may assume, as we have just done, that the mass m is independent of the speed of the body. This is the case for all everyday speeds and even those of our rockets and satellites, but not for the speeds of some atomic particles, which approach the speed of light. According to the principle of relativity (Chapter 10), the mass of an object does increase with speed if the velocity is close to that of light, but this fact does not invalidate Newton's law if it is stated as in Eq. (3–12) rather than as in Eq. (3–9). Newton intuitively chose the more correct form.

3|5 WEIGHT

Weight was defined as the force of gravitational attraction. A force is measured by the acceleration it imparts to a mass m. For a body that is allowed to fall freely the acceleration of gravity is called $\mathbf{g}$. So the equation $\mathbf{F} = m\mathbf{a}$ tells us that

weight $= m\mathbf{g}$, \hfill 3–13

since it is a force that will give a mass m the acceleration $\mathbf{g}$.

In the mks system, weight is measured in newtons, since the newton is the unit for force. Thus for a mass of 1 kg near sea level, where $g = 9.8 \text{ m/sec}^2$, the pull of gravity or

the weight of 1 kg $= 1 \text{ kg} \times 9.8 \dfrac{\text{m}}{\text{sec}^2} = 9.8 \text{ N}.$

Similarly, near sea level the pull of gravity on 0.454 kg, the quantity of matter in a pound-weight, is 0.454×9.8 N. This particular force is the American and British engineering unit of force called "a force of one pound." Therefore

a force of one pound $= 0.454 \times 9.8 \text{ N} = 4.45 \text{ N},$

which makes it over four times as large a unit as the newton. A force of 2000 lb, or $2000 \times 4.45 \text{ N} = 8900 \text{ N}$, is called "a force of one ton."

Example 1. At blastoff an upward thrust (force) of 12 tons is applied to a rocket whose weight (at the earth's surface) is 2 tons. What is the initial acceleration upward?

Solution. The net upward thrust **F** is 10 tons, or five times the weight. When the only force acting on a body is its weight, the body experiences an acceleration (near the earth) equal to g, or 9.8 m/sec^2. A net force 5 times as great will impart an acceleration of $5g = 49 \text{ m/sec}^2$.

Example 2. How fast must an elevator accelerate downward if it is to support half of the weight of an occupant? (See Fig. 3–10.)

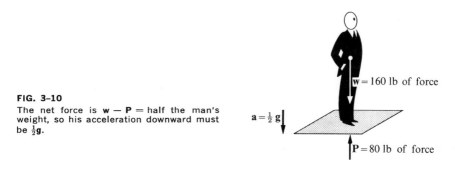

FIG. 3–10
The net force is $\mathbf{w} - \mathbf{P} =$ half the man's weight, so his acceleration downward must be $\tfrac{1}{2}\mathbf{g}$.

$\mathbf{a} = \tfrac{1}{2}\mathbf{g}$

$\mathbf{w} = 160$ lb of force

$\mathbf{P} = 80$ lb of force

Solution. Let m be the mass of the occupant. This person is subjected to two forces, namely (1) his weight $m\mathbf{g}$, a downward force, and (2) the upward push of the floor, which is given as $mg/2$ in magnitude. The net force **F** is

$\mathbf{F} = m\mathbf{g} - mg/2 = m\mathbf{g}/2$ down.

This net force will result in a downward acceleration **a** half as great as that due to weight alone, so $\mathbf{a} = \mathbf{g}/2$. If the person is to have this acceleration, the elevator must also. Were the occupant standing on spring scales in the elevator, the scales would be supporting half of his weight and would read accordingly.

3|6 NEWTON'S FIRST LAW OF MOTION. EQUILIBRIUM

It follows from Newton's second law of motion that if the net force **F** acting on a body is zero, then

$$\mathbf{F} = \frac{\Delta\boldsymbol{\mu}}{\Delta t} = m\mathbf{a} = 0,$$

or the momentum of the body will not change, its velocity will remain constant, and its acceleration will be zero. We thus have the following theorem:

A body subject to no net force will continue in motion at constant velocity or, if at rest, will continue to be at rest.

This is known as *Newton's first law of motion.* We see that while it is a special case of Fundamental Law I, it specifies just what we mean by "no force," or "zero force." We mean no *net* force, rather than no applied force. Newton's first law recognizes the existence of friction in natural motions and reminds us that it must be taken into account. Even on a smooth horizontal frozen pond a moving ice puck will not keep going indefinitely because of some friction and air resistance. For constant velocity these retarding forces must be balanced by an additional force due to an agent, gravity, or some other cause.

If the forces applied to a body act at the same point and the forces in any direction equal the forces in the opposite direction, then the net force **F** is zero, and the above theorem applies. The force **F** will also be zero if the applied forces added vectorially form a closed polygon. If $\mathbf{F} = 0$ and the body is already at rest, then it will remain so and is said to be in a state of *equilibrium.*

Example. A boy sits in a swing supported by a long rope. A man pulls the boy out until the rope makes a considerable angle with the vertical and then holds the boy at rest in a state of equilibrium. How do the forces acting on the boy balance?

64

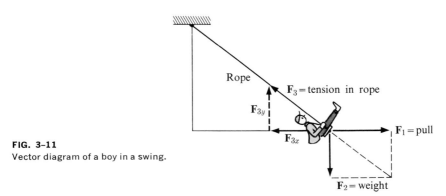

FIG. 3–11
Vector diagram of a boy in a swing.

Solution. In Fig. 3–11 the pull $\mathbf{F}_1$ is taken to be to the right. The boy's weight is the downward force $\mathbf{F}_2$. The third force applied to the boy is the tension $\mathbf{F}_3$ in the rope. For equilibrium the horizontal projection, or component, of $\mathbf{F}_3$ to the left, that is, $\mathbf{F}_{3x}$, must balance $\mathbf{F}_1$, and the upward component of $\mathbf{F}_3$, $\mathbf{F}_{3y}$, must balance $\mathbf{F}_2$. Thus for a given pull and weight, the direction and magnitude of $\mathbf{F}_3$ are determined.

3|7 FORCE AND MOTION

We are now ready to consider some of the more important types of motion that result under various kinds of net force. We shall limit ourselves to the motion of a small particle of finite mass, or to the motion of the center of an object such as a ball, omitting any discussion of the more complicated rotational motions that are possible for extended bodies. Let us assume that we may neglect the relativistic effects that are important only at speeds close to that of light, so that we may take

$$\mathbf{F} = m\mathbf{a}$$

as our fundamental equation relating force and motion.

a | $\mathbf{F} = \mathbf{0}$.

This case has just been discussed. The only possible motion is one at constant velocity. The distance s that an object with speed v travels in a time t is, according to the definition of velocity,

$$s = vt.$$

b | **F** *is constant.*

A constant net force results in a constant acceleration. If the object is accelerated from rest, it will travel in a straight line in the direction of the force, which is the case discussed in Section 3–1 in connection with Galileo's experiments. Then the distance traveled from rest in the time t is

$$s = \tfrac{1}{2}at^2.$$

When a ball is thrown obliquely upward its motion may be broken down into a combination of a vertical and a horizontal motion. Let us neglect air resistance. Then the only force acting is the constant pull of gravity, $m\mathbf{g}$, downward. Since there is no horizontal force, the horizontal motion falls under class (a) above and the ball moves horizontally equal distances in equal times. The vertical motion falls under class (b) above, except that the ball is given an initial upward velocity opposite to the direction of the constant force. As a result of the constant acceleration downward, the ball's upward velocity steadily diminishes to zero, and then the ball falls with an increasing downward velocity. This vertical rise and fall of the ball, combined with its steady horizontal motion, result in the parabolic path that balls, water jets, and projectiles are observed to follow when given an oblique initial velocity.

c | **F varies with the speed.**

This is the case for frictional forces, which are always present to some extent. As an example, suppose that the net force $\mathbf{F}$ has the magnitude

$$F = F_1 - F_2,$$

where $\mathbf{F}_1$ is a constant forward force and $\mathbf{F}_2$ the opposing frictional drag, which we shall assume to be proportional to the speed. Then $F_2 = kv$, where k is a constant. In this case, $F = ma$ becomes

$$F_1 - kv = ma, \qquad a = \frac{F_1 - kv}{m}.$$

We see that if a is positive it will decrease as the speed v increases until $kv = F_1$, when the acceleration will become zero and the body will attain what is called the *limiting velocity* v_l, for which $kv_l = F_1$. This is what happens when a parachutist falls ($F_1 = mg$) against air resistance. It is also the case where a constant force is applied to pull or propel a vehicle, such as a car or a cart.

d | **F = −ks, where k is a constant.**

Here we have a force of restitution opposite and proportional to the displacement $\mathbf{s}$. This type of force occurs when a spring is stretched or a pendulum bob is started swinging. As the object moves out from its natural position of equilibrium, the force of restitution brings it to rest at the extreme point of its swing, after which the object gathers momentum in the opposite direction, shoots past its equilibrium position, is again slowed to rest by the reversed force **F**, returning to repeat the cycle periodically. For an object of mass m the displacement varies sinusoidally with the time, that is,

$$s = A \sin (\sqrt{k/m}\ t)$$

and the motion is called *simple harmonic*. Here A is the amplitude of the swing, since the sine function ranges from $+1$ to -1 (see Appendix 4).

These examples all illustrate the application of our Fundamental Law I. They pertain to a branch of mechanics called *dynamics*.

3|8 NEWTON'S THIRD LAW OF MOTION. FUNDAMENTAL LAW II

To complete his theory of mechanics, Newton found that he needed another fundamental postulate which is quite independent of his other laws. This new law is frequently called *Newton's third law of motion*, but we shall refer to it as *Newton's action-reaction law*, since this title will always remind us of the subject with which this law is concerned. The importance, in fact the necessity of this law in mechanical theory is not always apparent, as it is frequently tacitly assumed, but illustrations will be given that should clarify its meaning. This law may be stated as follows.

When two objects interact, the force exerted by the first on the second (the action) is equal in magnitude and opposite in direction to the force exerted by the second on the first (the reaction).

If we label the action $\mathbf{F}_{12}$ and the reaction $\mathbf{F}_{21}$, then the law may be expressed mathematically as

II. $\mathbf{F}_{12} = -\mathbf{F}_{21}.$ 3–14

This is our *Fundamental Law II*.

To illustrate the action-reaction law let us consider some familiar examples.

Example 1. *A pushing force.* If a person pushes his finger down against the top of a desk with a force of magnitude F, the desk will push up against his finger with a force of the same magnitude.

Example 2. *A pulling force.* If one pulls out on a hook in the wall, he will feel a force on his arm toward the wall. The strength of the action and reaction may be measured by hooking the bottom ends of two spring balances together, attaching the top of one spring balance to the hook and grasping the top of the other in one hand (Fig. 3–12). If you try this experiment you will see that the two spring balances read the same. You should next disconnect the first spring from the wall and let a friend hold its far end so that you are pulling against one another. Again the readings will be the same.

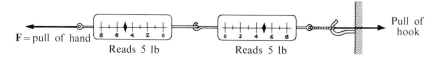

F = pull of hand   Reads 5 lb   Reads 5 lb   Pull of hook

FIG. 3–12
Action and reaction measured by two spring balances.

Example 3. *A supporting force.* Suppose that one holds a child in his arms. He will feel the child's body pushing downward against him. The child, however, will feel an upward lifting force exerted by the person holding him.

Example 4. *Gravitational attraction.* The earth attracts the moon and the moon attracts the earth. The pull of the moon on the earth gives rise to the tides. Similarly, a falling apple is pulled down by the earth's gravitational attraction, and at the same time the earth is attracted toward the apple (Fig. 3–13). To an outside observer the apple and earth would appear to move toward each other, but because of its much greater mass (inertia) the earth would be seen to gain speed *much* more slowly than would the apple.

The above examples all illustrate one important aspect of the action-reaction law, namely that *the action and corresponding reaction never act on*

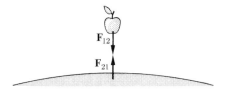

F_{12}

F_{21}

FIG. 3–13
An apple falling freely toward the earth experiences a force F_{12} exerted by the earth, and exerts an equal and opposite force F_{21} on the earth.

the same body. For this reason a force acting on a body is not nullified by the reaction to this force, since the reaction does not act on the body in question. Thus in Example 4 the apple is only acted on by the pull of the earth, and so the apple is not in equilibrium but is being accelerated. A body can be in equilibrium only if all the forces acting on the body add up to zero.

Another important point to note is that an action cannot exist without a reaction. Action and reaction always appear as a pair of forces. If one disappears, the other does also. This means that an object cannot exert a force unless there is some other object present upon which this force may be exerted, for without the other object a reaction could not arise. One cannot push against empty space! If one pushes or pulls against a fixed object such as the floor or wall, the reaction certainly continues as long as does the action. But what will happen if one pulls an easily moved object such as a toy wagon? In this case the force of the pull may easily exceed the retarding force of friction, and so there will be a net forward force on the wagon which will cause it to be accelerated. The wagon will soon acquire a velocity equal to that with which one wishes (or is able) to pull it, and as it does so the person pulling the wagon will unconsciously reduce the force of his pull until it is just enough to balance the retarding force. Throughout the process he will experience the reaction of the wagon against his hand, but this reaction will also vary, being greatest while the wagon is accelerating and much less when its speed is constant. In other words, the reaction will at each moment be equal and opposite to the instantaneous value of the action.

If one pushes against a wall, he will slightly deform the wall, causing a large number of its atoms to be crowded ever so slightly closer together. The displaced atoms will repel each other strongly (another example of action and reaction). Only the atoms touching his finger will exert the reaction against him, but the effect of his pushing action is transmitted into the wall through the internal interactions between its atoms.

Since action and reaction occur together as a pair of forces, should one of the forces in such a pair always be labeled the "action" and the other the "reaction"? The answer to this question is "No." When the forces are due to interactions between inanimate objects, such as molecular or planetary bodies, neither $\mathbf{F}_{12}$ nor $\mathbf{F}_{21}$ in Eq. (3–14) is more the action or the reaction than is the other. The force $\mathbf{F}_{21}$ is the reaction to $\mathbf{F}_{12}$, and $\mathbf{F}_{12}$ is the reaction to $\mathbf{F}_{21}$. However, when one of the forces is exerted by an animate being as the result of a mental decision by this being, then it seems natural to call this force the action, even though this distinction carries no physical significance.

3|9 THE PHYSICAL EXPLANATION OF LOCOMOTION

Locomotion furnishes many interesting, even surprising, examples of the action-reaction law.

Example 1. *Walking.* While a person stands on the floor, the only external forces acting on his body are (1) a downward force *mg*, called his weight, which is due to the earth's gravitational pull, and (2) the force exerted by the floor against his feet (Fig. 3–14). This second force is *not* the reaction to his weight; rather it is the reaction to the force which he exerts through his feet against the floor. If he is standing still, the net force acting on him must be zero, and so the two forces acting on him then actually are equal and opposite. However, when he commences to walk, he is being accelerated forward, which means that a net external force in the forward direction must act on his body. This forward force is supplied by the floor, whose total action on him is now both upward and forward. To make the floor supply a forward push, one must push backward on the floor. One does this with his feet, counting on friction (here it is *static friction*) to prevent his feet from slipping. It is, then, through the friction between the feet and the floor that one obtains the desired forward force. To realize the value of friction just imagine the difficulty of walking on very smooth ice.

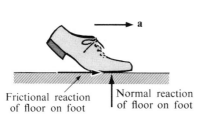

Frictional reaction of floor on foot | Normal reaction of floor on foot

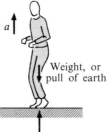

Weight, or pull of earth

Upward reaction of floor

FIG. 3–14
Reactions exerted by a floor on the shoe of a man walking.

FIG. 3–15
Man jumping: the upward reaction of the floor exceeds his weight.

Example 2. *Jumping.* Since jumping involves an initial upward acceleration it can only occur when there is a net upward force acting on the jumper. In the previous example weight was balanced by the upward reaction of the floor. How could we ensure that the upward push of the floor would exceed our weight?

It was pointed out that the push of the floor on a person is not the reaction to his weight, but to the force exerted by his feet against the floor. Therefore, to cause the push of the floor on him to increase, he must increase his push against the floor; this is what we do when we bend our knees and then suddenly straighten them as we jump. The action and reaction between our feet and the floor now exceed our weight in magnitude (Fig. 3–15).

Example 3. *Traveling by car.* Let us consider only the external forces that act on an automobile as it accelerates. These forces (see Fig. 3–16) are the following: (1) the weight, acting down, (2) the upward force exerted by the road on the tires, (3) a backward drag due to air and road resistance, and (4) a forward horizontal force exerted by the road on the rear wheels. If the car is accelerating, the fourth force must exceed the third. Through the action of the motor, the wheels are caused to push back harder on the road, thus causing the road to supply the required forward thrust to the car. Here again the action and reaction are of a frictional nature and so dependent on the nature of the surfaces in contact.

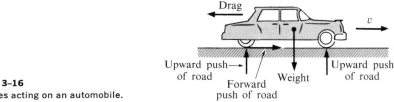

FIG. 3–16
Forces acting on an automobile.

In the case of a wheel rolling on soft ground, the ground becomes depressed under the wheel, and the wheel is continually trying to roll up out of the depression. In this case the ground, at the forward point of contact with the wheel, exerts a *backward* force on the wheel and hence on the car. This force must not be confused with frictional force that opposes skidding and pushes the car forward. The backward force on a rolling wheel is simply an additional drag on the car; note that it was referred to in (3) as the "road resistance."

Example 4. *Traveling by boat.* The forward force acting on a boat is supplied by the water against the blades of the oars or propeller. The harder an oar pushes against the water in a backward direction, the harder the water pushes forward on the oar, which is part of the boat.

71

Example 5. *Flying.* An airplane is acted on by four external forces (Fig. 3–17), as follows: the weight $\mathbf{F}_1$ down, the lift $\mathbf{F}_2$ up, the drag $\mathbf{F}_3$ backward, and the thrust $\mathbf{F}_4$ forward. To each of these there is an equal and opposite reaction exerted *by the plane* on its surroundings. Since $\mathbf{F}_1$ is exerted on the plane by the earth, its reaction is the gravitational pull of the plane on the earth. The reaction to $\mathbf{F}_2$ is a downward push against the air, the reaction to $\mathbf{F}_3$ is a forward drag on the air, and the reaction to $\mathbf{F}_4$ is a backward push against the air.

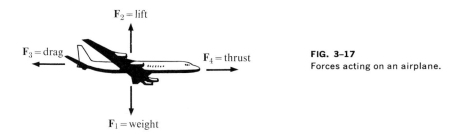

FIG. 3–17
Forces acting on an airplane.

If the plane is in equilibrium, that is, if it is flying horizontally at constant velocity, then $\mathbf{F}_2 = -\mathbf{F}_1$ and $\mathbf{F}_4 = -\mathbf{F}_3$. But the plane may be climbing, in which case $\mathbf{F}_2$ must exceed $\mathbf{F}_1$ in magnitude, or the plane may be accelerating, in which case $\mathbf{F}_4$ must exceed $\mathbf{F}_3$. Since action and reaction are always equal and opposite and act on different bodies, $\mathbf{F}_1$ obviously cannot be the reaction to $\mathbf{F}_2$, or $\mathbf{F}_3$ the reaction to $\mathbf{F}_4$.

As man's progress in flying has carried him to higher altitudes he has found the air to be less and less dense. Since one cannot push against empty space, one cannot expect to obtain a forward thrust from it. Therefore, man has resorted to jet engines in which gases are pushed out backwards, thus resulting in a forward thrust on the vehicle by the expelled gas.

Example 6. *Space travel.* Above the earth's atmosphere a vehicle will be acted on by the gravitational forces due to the heavenly bodies, of which only one or two may have a significant effect. The rocket (jet) principle may be used to create additional forces, but fuel supplies are limited on a prolonged space trip, and so rockets must be used mainly for guidance and for takeoff and landing. If only the net gravitational force acts, it will cause a spaceship to be accelerated in the direction of that net force. We shall see in the next chapter that this acceleration is welcome if it is just what is needed to keep the vehicle in the desired orbit, as is the case for earth-circling satellites.

3|10 COLLISIONS AND EXPLOSIONS. THE CONSERVATION OF MOMENTUM

When two objects A and B meet in a collision they interact, that is, A exerts a force $\mathbf{F}_{AB}$ on B, and B exerts a force $\mathbf{F}_{BA}$ on A. Each of these forces may vary with time, but at any instant they must, according to the action-reaction law, be equal and opposite, so that

$$\mathbf{F}_{AB} = -\mathbf{F}_{BA}. \qquad\qquad \textbf{3-15}$$

These two forces exist simultaneously for *equal* intervals of time. They may arise only at the moment of physical contact and last for a fraction of a second, as when two billard balls collide, or they may originate when the approaching objects are far apart and exert gravitational or electrical forces on each other for a long period of time, as in the case of an astronomical collision. According to Newton's law of motion, Eq. (3–12), the change in the momentum of a body when a force $\mathbf{F}$ acts on it for a time Δt is

$$\Delta\boldsymbol{\mu} = \mathbf{F}\,\Delta t.$$

Since $\mathbf{F}_{AB}$ and $\mathbf{F}_{BA}$ have equal magnitudes and act for the same time they must produce changes in momentum of equal magnitudes. However, since $\mathbf{F}_{AB}$ and $\mathbf{F}_{BA}$ are opposite in direction, the resulting momentum changes must also be in opposite directions. Finally, since $\mathbf{F}_{AB}$ acts on B and $\mathbf{F}_{BA}$ acts on A, one of the momentum changes is that of B and the other that of A. Whatever momentum B gains in, say, the positive x-direction, A gains in the negative x-direction. A gain in the negative x-direction is equivalent to a loss in the positive x-direction, so whatever momentum one object gains the other loses. Hence in any collision momentum is passed from one body to another, but it is neither created nor destroyed; in other words, *momentum is conserved in a collision*.

What has been said about collisions also applies to explosions. Suppose that through the release of stored energy a body breaks into two fragments A and B. While A and B are being torn apart and accelerated they must exert equal and opposite forces, $\mathbf{F}_{AB}$ and $\mathbf{F}_{BA}$, on each other, and these forces must exist for equal times. The reasoning above then tells us that here too

$$\Delta\boldsymbol{\mu}_A = -\Delta\boldsymbol{\mu}_B, \qquad \Delta\boldsymbol{\mu}_A + \Delta\boldsymbol{\mu}_B = 0. \qquad\qquad \textbf{3-16}$$

Nor does it matter how many fragments are produced; the sum of the momentum changes will be zero.

A word of caution about signs is in order here. Momentum is a *vector* quantity, having the direction of the associated velocity. In one-dimensional problems, we may choose momentum in one direction, say to the right, to be positive, and then we must call momentum in the opposite direction (left) negative.

In a collision or explosion, we may regard all of the interacting objects (A, B, etc.) as the parts of a *system*. The forces $\mathbf{F}_{AB}$ and $\mathbf{F}_{BA}$ may then be considered as internal interactions within the system. We have seen that these forces alone cannot change the momentum of the system. Therefore, if no external forces act on a system, its momentum will remain constant. Finally, if external forces act on a system but the sum (resultant) of these forces is zero, the momentum of the system will be conserved. This theorem is called the *conservation of momentum principle;* it is a derived law, derived from our first two fundamental laws, the law of motion and the action-reaction law.

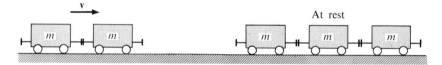

FIG. 3–18
A coupling collision.

Example 1. Two coupled freight cars, each of mass m, are moving with an initial speed v. The two cars collide and couple with three stationary cars, each of mass m (Fig. 3–18). What is the final speed of the five-car train?

Solution.   Let v' be the final speed.

Momentum before collision $= 2mv$.
Momentum after collision $= 5mv'$.
Momentum after collision $=$ momentum before;

therefore,
$$5mv' = 2mv,$$
$$v' = 0.4v.$$

So, if the two cars were traveling at 20 mi/hr, the five would move along at only 8 mi/hr.

Example 2. A 1-ton sports car runs head-on into a 5-ton truck. The truck driver survives and testifies that he was going 32 mi/hr before the accident and that after the collision he initially traveled at 10 mi/hr with the wrecked car on his hood. How fast was the sports car going?

Solution. Take the direction of motion of the sports car as positive, that of the truck as negative. Let v be the initial speed of the sports car.

Momentum before collision $= [(1)v + 5(-32)]$ ton-mi/hr.

Momentum after collision $= [(6) \times (-10) = -60]$ ton-mi/hr.

Momentum before collision = momentum after.

$$v - 160 = -60$$
$$v = 100 \text{ mi/hr.}$$

If an object of mass m is whirling in a circle of radius r with a speed v, the product mvr is called its *angular momentum*. Considerations based on Newton's laws (and similar to those discussed in connection with linear momentum) lead to the conclusion that if a rotating system has little connection, through frictional or other outside forces, with its surroundings, *its angular momentum must be conserved*. Thus if the man in Fig. 3–19, who is rotating on a platform supported by ball bearings, pulls in his arms and the weights he is holding, he will spin faster. If the product mvr is to remain constant, a decrease in r must be accompanied by an increase in v.

FIG. 3–19
Conservation of angular momentum.

The conservation of angular momentum theorem finds frequent application in nuclear physics in connection with the disintegration of a spinning particle into two or more other particles; the angular momentum of the original particle is passed on to its decay products. Newton's laws of motion appear to hold good throughout the world, from the subatomic to the astronomical level.

PROBLEMS   1.   A body falls from rest with a constant acceleration of 9.8 m/sec^2. Find

    a)   the speed at the end of each of the first five seconds,

    b)   the average speed during each of the first five seconds,

    c)   the distance traveled during each second, and

    d)   the distance fallen from rest after 1, 2, 3, 4, and 5 sec.

2.   A net force **F** is found to give an acceleration of 6 m/sec^2 to a 5-kg object and an acceleration of 2.4 m/sec^2 to an object of mass m'.

    a)   Find **F** and m'.

    b)   What definition did you use in each calculation?

    c)   Did you assume Newton's law of motion and, if so, how?

3.   A net force of 0.4 N is found to give a mass m an acceleration of 0.12 m/sec^2, while a net force **F'** gives the same mass an acceleration of 0.30 m/sec^2.

    a)   Find **F'** and m.

    b)   Did you assume Newton's law of motion and, if so, how?

4.   The pull of gravity varies inversely as the square of the distance from the *center* of the earth.   Take $g = 9.8 \text{ m/sec}^2$ at sea level, for which the earth's radius is about 4000 mi.   Determine the weight in newtons of a 75-kg man

    a)   at sea level,

    b)   on Mt. McKinley (4 mi high), and

    c)   at an altitude of 4000 mi.

5.   Add graphically to find the sum of the following forces: $\mathbf{F}_1 = 100 \text{ N}$ to the right, $\mathbf{F}_2 = 200 \text{ N}$ up, and $\mathbf{F}_3 = 300 \text{ N}$ to the left and down at an angle of $45°$.

6.   A 400-gm block rests on a horizontal table.   Find the acceleration under applied forces of 1, 2, 3, and 4 N, if in each case the frictional drag is 1 N.

7.   What opposing force of magnitude F will bring a body of mass m to rest in a time Δt if the initial speed of the body is v?   Compute F in newtons and pounds of force for a 2000-kg car going 29 m/sec (about 60 mi/hr) which must be stopped in 4 sec.

8. Show why a sailboat can follow a course that is somewhat (say 45°) upwind. Assume that the net force is perpendicular to the sail and that a centerboard or keel prevents a sidewise drift of the boat.

9. Show that $a/g = F/w$, where a is the acceleration of a body due to a net force F and w is its weight.

10. An 80-kg man stands on platform scales in an elevator. What will the scales calibrated in newtons read if the acceleration of the elevator is

    a)   zero,

    b)   0.98 m/sec² up, and

    c)   0.98 m/sec² down?

11. A weight is hung from a support by a thread. A second weight is hung by a thread tied to the bottom of the first weight. Explain how one may, by pulling down on the lower weight, first break either

    a)   the upper thread, or

    b)   the lower thread.

12. Explain why one may place his hand on a table, put a 10-kg iron block on his hand, and then hit the block a hammer blow from above without injury to the hand.

13. In a tug-of-war two boys pull on opposite ends of a rope and each exerts a force of 100 N (about 22.5 lb-weight).

    a)   What horizontal forces act on each boy?

    b)   What horizontal forces act on the rope?

    c)   Which of the above forces are action-reaction pairs?

14. A piece cut from the rope in problem 13 is hung vertically and made to support an increasing weight until it breaks. If the rope breaks when the mass tied to the end is 20 kg or greater, what is the breaking strength of the rope in newtons? Will the boys in problem 13 break the rope?

15. A skier climbs up a slope and then glides down it.

    a)   When is the frictional force of the snow on the skier directed up the slope and when is it directed down?

    b)   When is the frictional force on the skier in the direction in which he is going, and when is it in the direction opposite to his motion?

    c)   Must the frictional force be greater when going up or when going down and why?

77

16. a) Find the average recoil force on a machine gun that is firing 150 12-gm bullets per minute, each with a speed of 900 m/sec.

b) How long would it take these bullets to stop a 30-kg panther that springs toward the gunner with an initial speed of 5 m/sec?

17. a) A freight car of mass m moving at 20 mi/hr catches and couples with two similar cars that were moving at 5 mi/hr. Find the speed of the three cars just after coupling.

b) Repeat for the case where the cars have the same original speeds, but are approaching one another.

18. A projectile whose mass is 50 kg is fired directly forward by an airplane whose mass without the projectile is 5000 kg. If the plane's original velocity is horizontal at 480 mi/hr and that of the projectile is 840 mi/hr, both relative to the earth, find the plane's speed after firing the projectile.

19. Two men, each of mass m, sit facing each other in opposite ends of a boat of mass $2m$. One man throws a ball of mass $m/40$ to the other man, who catches it. If the horizontal velocity of the ball is 80 ft/sec, what will be the velocity of the boat

a) while the ball is in the air,

b) after it is caught? (Neglect water and air resistance.)

20. Suppose that a rocket emits 200 kg of exhaust each second from its rear and that the velocity of the exhaust relative to the rocket is 2500 m/sec

a) What thrust in newtons will be developed between the rocket and its exhaust?

b) Express this thrust in pounds of force.

c) How much speed will the rocket gain per second when its mass is 20,000 kg and it is traveling vertically upward? [*Hint:* Do not neglect the weight.]

Experiment 5   THE INVESTIGATION OF FREE FALL
Aristotle versus Galileo

Object: To determine experimentally how the time it takes a body to fall depends on (1) the mass or weight, (2) the material, and (3) the distance fallen.

Theory: Aristotle argued that since a large stone is heavier than a small one, the large stone will fall faster. Galileo claimed that experimentally a large and a small stone fell at almost the same rate. You are to determine which of these famous men was right, and then you are to investigate free fall as thoroughly as possible. Before stating a conclusion, you must consider the possible error in your measurements. For example, if the time of fall of one object is measured as 1.2 ± 0.2 sec and that of a second object as 1.3 ± 0.2 sec, then you must conclude that within the limits of your measurements no difference in rates of fall was *proved* (though it may have been suggested). Two measurements such as 1.0 ± 0.1 and 1.3 ± 0.1 are definitely different. Thus you see that you should repeat each experiment several times (say five times) so as to obtain the possible error. Of course good timing technique should be developed to reduce your errors.

Procedure

Step 1. In five trials find the time it takes a small stone or 500-gm weight to drop a given distance of at least 15 ft. Repeat for a larger stone or weight. Was Aristotle or Galileo right?

Step 2. In five trials find the time it takes a pingpong ball or a ball of light paper to fall the same distance. Now who was right, Aristotle or Galileo? How would you explain the results of Step 1 and Step 2?

Step 3. Time how long it takes the heavy stone or weight to fall through two other distances, and measure all three vertical distances that you used. Call the distance s and the time t. Plot whichever of the following will give a straight line graph: s against t, s against t^2, s against $\sqrt{t}$, s against $1/t$, etc. How does s depend on t? What is the constant of proportionality? From its value compute g, the acceleration of a freely falling body.

THE FORCE TABLE   **Experiment 6**
Vector Addition and Static Equilibrium

Object: To observe how vectors add and to verify the conditions for static equilibrium.

Theory: If two or more vectors are laid off end to end, each with its proper magnitude and direction, then the vector connecting the beginning of the first with the end of the last vector is, by *definition*, the sum of the vectors considered. If the sum is *zero*, the vectors when added should form a *closed polygon*.

According to Newton's first law, if an object such as a ring is stationary (in equilibrium, we say), then the sum of the forces acting on the ring must be zero. When added graphically the forces should then give a closed polygon. Any failure to do so must be attributed to experimental error.

What errors may arise? First there are errors in finding a set of forces that will experimentally produce equilibrium. Due to friction, some variation in one force is possible without producing a noticeable effect. This source of error must be investigated. Second, errors may arise in the plotting of the polygon with ruler and compass; such errors may be minimized with care.

Procedure

Step 1. To find the error due to friction, put a 500-gm weight on one string and a 500-gm weight on another opposite string. Find how much extra force is needed to move the ring noticeably when the table is tapped as weights are added to one side. The largest amount you can add without detectable motion of the ring is the instrumental error per 1000 gm due to friction.

Step 2. Set two 500-gm weights at 120° apart. Find the third force (magnitude and direction) that will give equilibrium.

Step 3. Find the force to balance a 300-gm weight at 0°, a 400-gm weight at 90°.

Step 4. Find the force to balance a 200-gm weight at 0°, a 100-gm weight at 90°, and a 100-gm weight at 180°.

Step 5. Add the forces in Step 2 by the ruler and compass method. Do you get a gap rather than a closed polygon? If so, measure the gap and compute what force it represents. Try to justify this experimental error. Repeat for each of the other two sets of forces (Steps 3 and 4).

Experiment 7   COLLIDING CARS
                     Newton's Laws—Meaning of Mass

Object: To verify Newton's law of motion and action-reaction law and to see that the property of a body known as *inertia* is measured by its *mass*.

Theory: If two cars are connected by a stretched rubber band, then according to Newton's action-reaction law, the elastic must exert equal and opposite forces F_1 and F_2 on the two cars. If the cars have masses of m_1 and m_2, respectively, then Newton's second law tells us that the accelerations should have the magnitudes

$$a_1 = F_1/m_1 \quad \text{and} \quad a_2 = F_2/m_2,$$

so that if $F_1 = F_2$, we should find that

$$a_2/a_2 = m_2/m_1.$$

Now since distance traveled from rest under a given acceleration is $s = \frac{1}{2}at^2$, in the same time the two cars should travel distances proportional to their respective accelerations, or inversely proportional to their respective masses. If we can verify this conclusion experimentally, then we shall have obtained experimental evidence supporting Newton's laws of motion.

Procedure

Step 1. Weigh separately the two cars. Connect them with several elastic bands tied in series. Using the meter stick, separate the front ends 100 cm and then release cars simultaneously. Record the distance traveled by each before they collide. Repeat four times, invalidating any run in which motion is hampered by the slack elastic or is not parallel to meter stick. Reverse cars and repeat five more times. (Why?)

Step 2. Repeat Step 1, first with 200 gm added to one car, then with 400 gm added. Do your results confirm Newton's laws within experimental error? Explain. What role did friction play?

4|1 WHAT KINDS OF FORCE EXIST IN NATURE?

We have seen that Galileo investigated various types of motion, notably motion at constant velocity and motion under constant acceleration, while Newton described the relation between the motion of a body and the net force producing the motion. The next question to arise concerned the origin of forces. Where do forces come from? Newton gave a partial answer to this question with his theory of gravitation.

The most common forces in everyday life are pushes and pulls and gravitational attraction. The explanation of pushes and pulls of muscular origin is rather complicated and involves an understanding of muscular structure, chemistry, and atomic physics. However, the

Newton's Law of Gravitation

physicist believes that ultimately such forces may be attributed to forces between atoms which in turn may be resolved into electromagnetic forces (see Chapter 8). The physicist is really concerned only with the basic or fundamental types of force that exist in the physical world. We shall see, in the course of the book, that at the present time he recognizes the following fundamental types (see Fig. 4–1) listed in order of increasing strength:

a) Gravitational forces between two masses (Section 4–5).

b) The so-called "weak" beta-decay forces involved in radioactive decay (Section 14–4).

c) Electromagnetic forces, which include the related
   1. magnetic forces between moving charges (Section 8–2), and
   2. electric forces between charges at rest (Section 7–3).

d) The very strong forces between nuclear particles (Section 14–3).

It was the first of these five types of forces that Newton originally discovered or postulated.

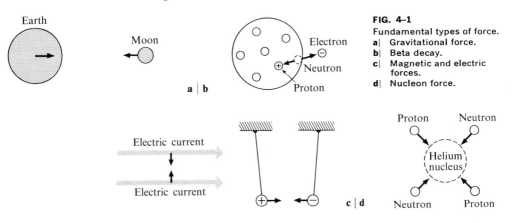

FIG. 4-1
Fundamental types of force.
a| Gravitational force.
b| Beta decay.
c| Magnetic and electric forces.
d| Nucleon force.

It should be borne in mind that we still do not know *why* these forces exist. We merely postulate that they do exist and show that we may then explain varied phenomena in terms of them. It is possible that two or more of these forces may be explained in terms of a single, still more basic, postulate. Einstein and Heisenberg have attempted to do this in their unified field theories, but we do not expect ever to find an ultimate explanation, since man is not omniscient. For our purposes the important thing is to reduce chaos to order, that is, to explain many things in terms of a few hypotheses. As the number of fundamental hypotheses is reduced through generalization, simplicity is gained up to a point, but beyond that point it appears that the increasing complexity of the concepts and mathematics makes the subject more difficult for the ordinary mind to grasp. Just when this point is reached is a matter of personal opinion and taste. This book merely reflects the author's (and others') opinion as to what is the clearest way to describe how the world around us behaves. Time will undoubtedly alter this viewpoint.

4|2 AN HISTORICAL SURVEY OF PLANETARY THEORIES

In Greece during the fourth century B.C., observations of the night sky led Plato and Aristotle to postulate a theory to explain the apparent motions of the heavenly bodies. It best suited the Greek philosophy, or metaphysics, to assume at the start that the earth is in the center of the universe and that other bodies move around the earth in orbits of the most "perfect" (to them) shape, namely, circles. These initial hypotheses led to the necessity of also postulating a most intricate system of revolving spheres. The whole theory was unable to lead to new discoveries and when further observations were made, the theory had to be patched up by being made still more complicated. One Greek, Aristarchus (third century B.C.), did propose a heliocentric (sun-centered) system, but it was not popular, and he gave no quantitative calculations. Ptolemy of Alexandria in the second century A.D. modified the Aristotelian theory by assuming motion in epicycles rather than in revolving spheres. Ptolemy postulated that a planet moved in a circle whose center moved in another circle and so on until one finally came to a circle whose center moved in a circle with the earth at its center (Fig. 4–2). This theory could be made to explain the observed facts, but it too was complicated and unfruitful.

The Polish scientist Copernicus (1473–1543) revived the heliocentric theory and carried out the quantitative calculations that were necessary in order to show that his theory could be made to explain the known facts. He found that he too had to postulate a complicated system of circular motions which, though a little simpler than that of Ptolemy, was no better at correlating the facts and opening up new discoveries. The importance of the work of Copernicus was that it (1) challenged the dogmatic belief in Aristotelian science, (2) showed that more than one theory could successfully explain a set of known facts, and (3) paved the way for Newton's theory of gravitation.

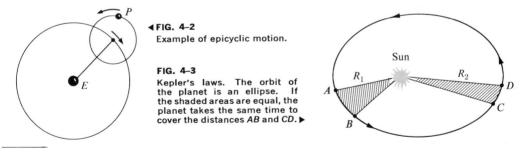

◀ **FIG. 4–2**
Example of epicyclic motion.

FIG. 4–3
Kepler's laws. The orbit of the planet is an ellipse. If the shaded areas are equal, the planet takes the same time to cover the distances *AB* and *CD*. ▶

Kepler (1571–1630) surveyed the more accurate observations of Tycho Brahe and found order in the data. Kepler described this order in his three empirical laws, which are as follows:

1. *Each planet moves in an elliptical orbit.*

2. *A planet moves in its orbit at such a varying speed that the line from the sun to the planet sweeps out equal areas in equal times* (Fig. 4–3).

3. *The square of the time of revolution (period) is directly proportional to the cube of the mean radius of the planet's orbit.*

While Kepler was carrying on his work in Germany, his friend and correspondent Galileo was eloquently defending the heliocentric theory in Italy. Galileo was the first to use the telescope to make astronomical measurements, and the results of his observations only added to the zeal with which he attacked the geocentric (earth-centered) theory and the scholastics who dogmatically defended that theory. He was eventually forced by the Inquisition to make a formal renunciation of the Copernican theory and to promise to discontinue making statements antagonistic to the Church. However, this did not change his beliefs.

Kepler's postulate of elliptical orbits made the heliocentric theory much simpler. Of course the "perfect" circle no longer played a part in the theory, but in place of several epicycles it was necessary to associate only a single ellipse with each planet. However, the theories of Ptolemy, Copernicus, and Kepler all correlated the observed facts and nothing more. Indeed we can still say today that the planets appear to move in the sky just as though their actual motions were those postulated by any one of the above theories. What has made Kepler's model the one now universally accepted by scientists is the work of Newton (1642–1727).

Newton *explained* Kepler's laws with his theory of universal gravitation. This theory took as its premises Newton's law of motion and a new fundamental law that Newton postulated, namely his law of gravitation. From these postulates Newton was able not only to derive Kepler's laws but also to explain many other things and to predict still others. We now say that it is gravitational attraction that keeps the planets from leaving the solar system, that keeps the moon in its orbit about the earth and Jupiter's moons moving around Jupiter, that makes objects fall toward the ground and gives us weight, and that accounts for the minute attraction observed between ordinary uncharged objects when a sensitive experiment is performed to measure this attraction. Without Newton's theory, we would not have artificial satellites, because no one would have thought them possible.

4|3 CENTRIPETAL ACCELERATION AND CENTRIPETAL FORCE

Let us imagine that we are looking down on a merry-go-round that is rotating counterclockwise at a steady rate (Fig. 4–4). Suppose that we notice a child riding on the eastern side at C. This child will be moving northward at the moment. We look away and happen to glance back at the merry-go-round after it has turned through half a revolution. The child is now at C' on the west side and is moving southward. How can a

86

velocity northward change into one southward? Well, perhaps another example will help us to answer this question.

Suppose that we are moving northward in a car and then apply the brakes, come to a stop, and begin to move backwards (southward), faster and faster. Our northward velocity was changed into a southward one through a *southward acceleration* produced by a southward force. So the child in the merry-go-round also must have experienced a southward acceleration due to a southward force even though the merry-go-round was turning steadily and the child's *speed* was constant. As the child swung around in a half circle from C to C' in Fig. 4–4, the center of the merry-go-round was, on the average, to the south of the child, so that the average acceleration was toward the center.

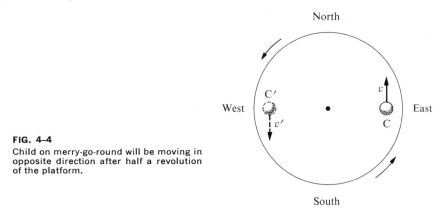

FIG. 4–4
Child on merry-go-round will be moving in opposite direction after half a revolution of the platform.

Newton showed (and we shall presently) that if an object is moving in a curved path, then the object is being accelerated toward the center of curvature of the path, regardless of whether or not the speed of the particle changes. This acceleration *toward* the center of a curved path is called *centripetal acceleration* $\mathbf{a}_c$ and we shall see that it arises because of the *change in the direction* of the body's motion. Remember that velocity is a vector quantity and that it may change in magnitude, or direction, or both. Change in velocity $\Delta\mathbf{v}$ in a time Δt implies the existence of an acceleration:

$$\mathbf{a} = \frac{\Delta\mathbf{v}}{\Delta t}.$$ 4–1

This is the definition of acceleration, Eq. (2–3).

Since a small arc of any curve may be regarded as an arc of a circle, let us consider a particle moving in a circle of radius r with a speed v.

87

At the point P in Fig. 4–5(a), the direction of motion will be along a tangent to the curve at P, so that the velocity $\mathbf{v}_1$ will be as shown. A short time Δt later the particle moving along the circle will reach Q, where its velocity $\mathbf{v}_2$ will be along the tangent to the curve at Q, as shown. We see that $\mathbf{v}_1$ and $\mathbf{v}_2$ have different directions. Let us assume that the speed does not change appreciably in the time Δt, so that $\mathbf{v}_1$ and $\mathbf{v}_2$ will have the same magnitude v.

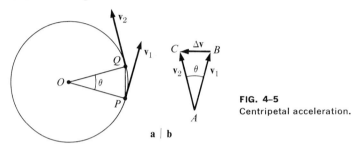

FIG. 4–5
Centripetal acceleration.

a | b

The *change* in a quantity is what one must add to the original value of the quantity in order to obtain its new value. In Fig. 4–5(b), $\mathbf{v}_1$ and $\mathbf{v}_2$ have been transposed, with their original directions unchanged, to the point A that serves as a common origin for the two vectors. Since a vector is specified by giving its magnitude and direction, but not its point of origin, $\mathbf{v}_1$ and $\mathbf{v}_2$ are the same vectors in Fig. 4–5(a) and (b). We see that the vector labeled $\Delta\mathbf{v}$ added end-to-end to the vector $\mathbf{v}_1$ gives the vector $\mathbf{v}_2$, so that

$$\mathbf{v}_1 + \Delta\mathbf{v} = \mathbf{v}_2,$$

$$\Delta\mathbf{v} = \mathbf{v}_2 - \mathbf{v}_1. \qquad\qquad 4\text{-}2$$

Since $\mathbf{v}_1$ is perpendicular to OP in Fig. 4–5(a) and $\mathbf{v}_2$ is perpendicular to OQ, the angle θ between OP and OQ is also the angle between $\mathbf{v}_1$ and $\mathbf{v}_2$. The triangles OPQ and ABC are similar triangles because they both have two equal sides enclosing the same angle θ. This means that the third side of triangle ABC, which is $\Delta\mathbf{v}$, must be perpendicular to the corresponding side PQ of triangle OPQ. A vector perpendicular to PQ at its midpoint will lie on a radius of the circle; therefore $\Delta\mathbf{v}$ must, for small values of θ, be directed *toward the center of the circle*.

In similar triangles corresponding sides are proportional; therefore

$$\frac{\Delta v}{v} = \frac{\overline{PQ}}{r}, \qquad \Delta v = \frac{v \cdot \overline{PQ}}{r}.$$

If θ is small, as we have assumed, $\overline{PQ}$ will very nearly equal the distance $v \, \Delta t$ that the particle moves along the arc of the circle in going from P to Q in the time Δt. If we set $\overline{PQ} = v \, \Delta t$, we have

$$\Delta v = \frac{v^2}{r} \Delta t,$$

$$a_c = \Delta v / \Delta t = v^2 / r. \hspace{3cm} \textbf{4-3}$$

This gives the magnitude of the centripetal acceleration $\mathbf{a}_c$. We saw that its direction is *toward* the center of the circular path.

From Newton's law of motion it follows that to give a body of mass m an acceleration $\mathbf{a}_c$ requires that a force $\mathbf{F}_c$, given by

$$\mathbf{F}_c = m\mathbf{a}_c,$$

must act on the body. Therefore, when a body of mass m moves with a speed v in a circular path of radius r, a force $\mathbf{F}_c$ whose magnitude is

$$F_c = mv^2 / r \hspace{3cm} \textbf{4-4}$$

and whose direction is *toward* the center of curvature of the path must act on the body. This force is called the *centripetal force*.

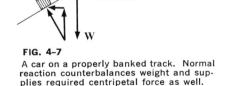

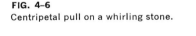

FIG. 4-6
Centripetal pull on a whirling stone.

FIG. 4-7
A car on a properly banked track. Normal reaction counterbalances weight and supplies required centripetal force as well.

There are many familiar examples of centripetal force, such as the following. (1) If a stone is whirled on the end of a string (Fig. 4-6), the string exerts a centripetal inward pull $\mathbf{F}_c$ on the stone; at the other end of the string the hand of the person doing the whirling pulls inwardly on the string. (2) When a car rounds a curve in the road (Fig. 4-7), the road must exert on the wheels of the car a force which is directed toward the inside of the curve; this force may be supplied by friction or by the normal reaction of a properly banked road. A road is properly banked for a given speed when a normal reaction alone will counterbalance the weight of the vehicle and supply the required centripetal force. Friction is a less dependable source of centripetal force since there is a limit (which varies with road and tire conditions) to how great friction can be. (3) A pilot flying an airplane in a turn banks it so that the air will exert the necessary sidewise centripetal thrust on the plane. (4) An electron circling the nucleus of an atom gets its inward centripetal force through the electrostatic attraction of the oppositely charged nucleus. (5) For a satellite the centripetal force is supplied by gravity, as we shall presently see.

4|4 CENTRIFUGAL FORCE

This is the reaction to the centripetal force. Since the latter is directed inward, centrifugal force must be directed outward. However, since action and reaction do not act on the same body, the centrifugal force does not act on the body following the curved path but on the object supplying the centripetal force. This point is often missed. Let us review the examples of the last paragraph.

(1) When a stone is whirled on the end of a string, the stone exerts an outward (centrifugal) pull on the string, which in turn exerts an outward pull $\mathbf{F}'_c$ on the hand of the operator. (2) When a car rounds a curve, the wheels of the car push down and outwardly against the road. (3) The circling plane pushes outwardly against the air. (4) The atomic electron attracts the nucleus with an outward centrifugal force. (5) The satellite exerts a gravitational pull on the planet around which the satellite is circling.

While granting all of the above, one may still wonder about the frequently expressed idea that centrifugal force tends to make the circling object fly outward and, indeed, will do so if the centripetal force is reduced. This is a misconception. If the centripetal force is reduced, the object will follow a less sharply curved path than before, but it will not fly *radially* outward. If the centripetal force is made zero, the object will travel out on a path *tangential* to its previously curved track. This follows from Newton's law of motion, for if the force on an object is zero, the acceleration must be zero and the velocity constant in speed and direction.

The impression that mud detached from a spinning wheel flies radially outward is false. However, if particles fly out tangentially from all sides of the wheel (Fig. 4–8), a spectator may get spattered whether he stands on one side or another and the resulting effect on his clothes may be just as disastrous as it would be if the mud were squirted out radially.

In a spin drier for a wet wash the rotational speed is increased past the point where the water can adhere to the rotating clothes. Then, like the mud that leaves the spinning wheel, the water flies off tangentially. However, it is caught by the sides of the drier, drains down to the bottom, and is thrown out.

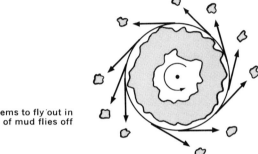

FIG. 4–8
Mud flung from a rotating wheel seems to fly out in all directions. Actually, each piece of mud flies off tangentially.

Centrifugal force is sometimes given quite a different definition. Suppose that we are inside of an enclosed merry-go-round which is turning so smoothly that we are oblivious of its motion. Objects that are not fastened down will not stay at rest relative to the floor, but will seem to move outward, like the water in the spin drier. Having forgotten that our floor is rotating, it is easy to believe that some outward "centrifugal force" is causing the objects to move outward. Actually such a force is purely *fictitious* and it is our reference system, and not the objects, that is accelerated. It is less confusing to avoid rotating frames of reference and to consider centrifugal force as a *real* force, an outward reaction on whatever produces a centripetal force.

4|5 NEWTON'S LAW OF GRAVITATION. FUNDAMENTAL LAW III

Newton developed the concept of, and formula for, centripetal force not long after he graduated from Trinity College in Cambridge, England. This led him to conclude that an inward force must act on each planet as it orbits the sun. It had been suspected for some time that the sun might attract the planets, and Newton adopted this view. In his famous book

Principia he proved mathematically that if a body moves in an elliptical orbit according to Kepler's second law (its radius vector sweeps out equal areas in equal times), then the body must be acted on by a force directed toward one of the foci of the ellipse. He further proved (after inventing the calculus to assist him) that for a body to follow an elliptical orbit, the force acting on it must be inversely proportional to the square of the distance from the attracting focus. Kepler's theory placed the sun at the focus of a planet's orbit, and hence Newton concluded that the sun exerted an inverse square force of attraction on a planet. He then boldly postulated that the attraction of a planet by the sun was just one example of a universal *law of gravitation* which states that:

Between any two objects in the world there exists a mutal force of attraction that is directly proportional to the product of the masses of the objects and inversely proportional to the square of their distance apart, or

$$F_g \propto \frac{m_1 m_2}{r^2},$$
4-5

where F_g is the magnitude of the force of gravitational attraction, m_1 and m_2 the respective masses of the attracting objects, and r their separation.

It has been assumed that the objects involved are small in size compared with their distance of separation. Newton proved that when the mass of a body is spherically distributed, the gravitational effect is the same as though all the mass were concentrated at the center of the body.

Let us define the gravitational constant G as the constant of proportionality in (4-5) when F_g, m_1, m_2 and r are measured in terms of previously defined units. We then have

III. $\quad F_g = G \dfrac{m_1 m_2}{r^2}$
4-6

as the mathematical formulation of our *Fundamental Law III*.

While Eq. (4-6) defines G, the postulate that G is a universal constant constitutes a fundamental principle or law of nature. We shall discuss its numerical value shortly.

It was characteristic of Newton's genius that he could make the inductive step of generalizing what he concluded to be the force between a planet and the sun by assuming that this same kind of force accounts for the pull of terrestrial objects toward the center of the earth and for the force that keeps the moon in its orbit around the earth.

Newton put his law of gravitation to both a qualitative and a quantitative test. He showed that by assuming this law one can deduce Kepler's third law which relates the period of revolution of a planet and the mean radius of its orbit. This derivation is given for circular orbits in Appendix 6. Newton also used his theory to calculate the period of the moon, which he was able to do without knowing the value of G; the computed period agreed very closely with the known value. Details of this computation will also be found in Appendix 6.

4|6 INERTIAL MASS VERSUS GRAVITATIONAL MASS

The concept of mass introduced in Chapter 3 associated mass with inertia; therefore the mass m in Newton's law of motion ($\mathbf{F} = m\mathbf{a}$) is more explicitly called the *inertial mass* of the body undergoing acceleration. The masses m_1 and m_2 in Eq. (4–6) are called the *gravitational masses* of the respective objects between which gravitational attraction exists. We shall follow Newton's example and assume that the gravitational mass is the same as the inertial mass of a body. In other words, Newton's law of gravitation postulates a relationship between previously defined quantities (force, mass, and distance). As a result, the weight of a body, which is the gravitational force on it, is proportional to its inertial mass. This in turn explains why Galileo was correct in saying that a light object and a heavy object should fall together with the same acceleration g provided that air resistance is negligible. For example, if the heavy object has twice the mass of the light one, then the downward force on the heavy object will also be doubled and the ratio F/m, which determines the resulting acceleration, will be the same for both objects.

The equivalence of inertial mass and gravitational mass is not, as far as we now know, logically bound to be true. It was an assumption on Newton's part and it is the cornerstone of Einstein's general theory of relativity. However, this postulate has been subjected to experimental test, and at present it has been verified that inertial mass and gravitational mass are the same to within one part in 10^9. This is really a remarkable result, for it means that the ability of a body to attract another one to it gravitationally increases linearly with the inertia of the body, or, stating it another way, both of these properties are proportional to the quantity of matter in the body. To convince yourself that this is not self-evident, note when we come to electrical forces (Chapter 7) that the ability of a charged body to attract or repel another one is *not* proportional to the mass of the body.

93

4|7 THE DETERMINATION OF G

Newton's tests of his theory did not involve knowing the numerical value of G. He left this for his successors to determine. Of course, he probably had a rough idea of the value of G, for it can be computed from g, the acceleration of free fall near the earth's surface, the radius R_E of the earth, and the mass of the earth (see Eq. 4–8). The trouble with this method of finding G was that the mass of the earth was then far from accurately established. One could only make a guess as to the average density (mass over volume ratio) of the earth and then multiply by the earth's volume $4\pi R_E^3/3$. But how could one make a good guess as to the earth's average density when one could then observe only its thin outermost crust?

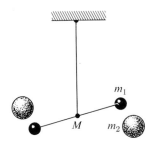

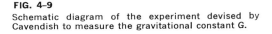

FIG. 4–9
Schematic diagram of the experiment devised by Cavendish to measure the gravitational constant G.

It was over 100 years after Newton published his *Principia* that another Englishman, Henry Cavendish (1731–1810), determined G accurately by directly measuring the gravitational force between *two terrestrial objects*. As shown in the schematic diagram of Fig. 4–9, the gravitational force between a pair of heavy balls (say 1.5 kg each) and a pair of lighter suspended balls is measured by determining the acceleration of each suspended sphere. Since the force is very weak, the restoring torque of the suspension as it twists must be small, and for this reason a fine quartz fibre is frequently used for the suspension. The very small acceleration of the suspended balls is magnified optically by reflecting a beam of light from a small mirror (attached at M) to a distant scale. A gravitational force of attraction is found to exist between each large and small ball, and one may verify the mass and distance dependence postulated by Newton. The constant of proportionality G is found to have a magnitude of 6.67×10^{-11} when F_g is expressed in newtons, m_1 and m_2 in kilograms, and r in meters (the mks units); G also has its units, which must be such as to make Gm_1m_2/r^2 equivalent to a force in newtons. This means that the mks units for G are the newton-meter^2/kilogram^2. We therefore write

$$G = 6.67 \times 10^{-11} \text{ N-m}^2/\text{kg}^2. \tag*{4-7}$$

It is because of the smallness of G that gravitational forces are not appreciable unless at least one of the attracting objects is of astronomical size. For two atomic particles the gravitational force is negligible.

Now that we know the value of G, we may use it to compute M_E, the mass of the earth. To do this consider an object of mass m on the surface of the earth (Fig. 4–10). Let R_E represent the earth's radius, whose value is well known from navigational measurements. The force of the earth's gravitational attraction on m is, according to the law of gravitation,

$$F_g = \frac{GmM_E}{R_E^2}.$$

FIG. 4–10
The weight *mg* of a terrestrial object is just the earth's gravitational attraction on it.

This force will impart the acceleration g to the mass m if the object is allowed to fall freely, so according to Newton's law of motion we may also write

$$F_g = mg.$$

These two expressions for the same force may be equated, so that we have

$$G\frac{mM_E}{R_E^2} = mg.$$

After canceling out the factor m (see Sec. 4–6) and solving for M_E, we get

$$M_E = \frac{g}{G} R_E^2, \tag*{4-8}$$

which expresses M_E in terms of known quantities. The numerical value of M_E is found to be

$$M_E = 6.0 \times 10^{24} \text{ kg.}$$

4|8 SATELLITES

With the advent of the space age Newton's theory has found spectacular applications of interest to everyone.   What is probably not generally known is that Newton himself thought about the possibility of creating an earth satellite and computed the speed it would need to have.   He discussed the firing of a cannonball from a very powerful cannon located on top of a high mountain (Fig. 4–11).   An ordinary projectile fired horizontally will gain downward velocity due to the pull of gravity and fall to earth in a curved path.   Fired at higher speeds, its path will curve less rapidly and its range will increase.   Suppose, thought Newton, that a projectile could be fired so fast that the curvature of its path matched that of the earth beneath it, then it would circle the earth back to the mountain from which it started.   The necessary speed turns out to be higher than a cannon could supply and a free satellite would maintain such a speed only if it were sufficiently high not to be slowed by the frictional effects of the earth's atmosphere.   The latter tapers off to where it is not a serious obstacle at an altitude of around 100 mi.   Let us now calculate the required speed.

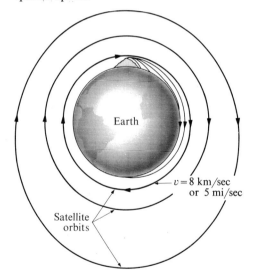

FIG. 4–11
Newton's hypothetical cannonball fired from a mountain at 5 mi/sec would, were it not for air resistance, circle the earth.

Suppose that a satellite of mass m circles the earth at an altitude h, so that the radius of its orbit is

$R = R_E + h.$

To make an object travel in a circular path a centripetal force mv^2/R must act on it. If the gravitational pull of the earth supplies just this force, then no other force will be needed and the satellite will circle freely. Therefore, we must have

$$\frac{mv^2}{R} = \frac{GmM_E}{R^2}.$$

A factor R, as well as m, may be canceled and the equation solved for v. We get (putting $R_E + h$ for R)

$$v = \sqrt{\frac{GM_E}{R_E + h}}. \hspace{3cm} \textbf{4-9}$$

The values of G and M_E were given in Section 4–7. The radius of the earth is about 6.4×10^6 m. Suppose, for example, that the height

$$h = 0.3 \times 10^6 \,\text{m} = 300 \,\text{km} = 186 \,\text{mi}.$$

Then

$$v = \left[\left(6.67 \times 10^{-11} \,\frac{\text{N-m}^2}{\text{kg}^2}\right) \times \frac{6.0 \times 10^{24} \,\text{kg}}{6.7 \times 10^6 \,\text{m}} \right]^{1/2} = 7.7 \times 10^3 \,\text{m/sec},$$

which is equivalent to about 17,400 mi/hr, or not quite *5 mi/sec*.

The period T, or time to circle the earth, is just the circumference of the orbit, $2\pi R$, divided by v. For our value of R,

$$T = \frac{2\pi \times 6.7 \times 10^6 \,\text{m}}{7.7 \times 10^3 \,\text{m/sec}} = 5460 \,\text{sec} = 91 \,\text{min}.$$

These figures have been publicized, but not so familiar is the fact that if an astronaut wishes to circle in a larger orbit, say one just above him, he must *accelerate* a little, so that he will spiral out, and then when he reaches the desired orbit he must *slow down* to a speed *smaller* than that which he had in his original orbit. Equation (4–9) states that the greater h the smaller v must be. The period increases with h, of course, until at the distance of the moon it is over 27 days.

Although physicists generally agree on their terminology in order that they may understand one another, when physics catches the public eye definitions may be somewhat distorted. A point in question is the term "weightlessness." A physicist would say that all objects near the earth, and this includes satellites and the astronauts in them, have weight, this being the gravitational attraction of the earth, which is independent of one's motion. However, astronauts in orbit are continually falling freely toward the earth, just as is a person who jumps off a high diving tower. In such circumstances the person is not supported, nor does he push against the floor or ground the way we earthbound creatures do. Put spring scales under his feet and, since they fall with him, the scales will read zero weight. So the term "weightlessness" has been applied to such situations. Astronauts in orbit cannot pour water, not because gravity has ceased to pull the water downward, but because as fast as the water falls the spaceship falls too and the water stays at a constant height above the ship's floor (Fig. 4–12). Although the situation is unfamiliar to us, the laws of Newton apply to it very well.

FIG. 4–12
In an orbiting spaceship, poured water forms a glob which hangs above (and does not fall into) a glass held beneath it. The water and the glass are falling together toward the earth.

4|9 THE CAUSE OF GRAVITATIONAL FORCES

Why do two bodies attract each other? Newton was unable to answer this question, but he pointed out that his theory did explain and summarize the known facts in mechanics in terms of the three postulates which we have labeled Fundamental Laws I, II, and III. This was a great simplification. His theory continued to explain new facts and to predict others and so it was a great success.

Of course Newton, and many since Newton, have speculated as to the cause of gravitation. One theory proposed that space was filled with a tenuous, invisible substance called the *ether* and that the action of the sun on the earth was transmitted through the ether, like pressure through the liquid in a hydraulic brake system. This theory was not fruitful; it did not grow gracefully to meet new facts and it has long since been abandoned.

Einstein's general theory of relativity takes a step toward explaining gravity by postulating that the accelerations we attribute to gravitational forces are simply the result of the curvature of space* in the neighborhood of a large mass. This theory predicted the observed bending of light rays that pass near the sun as well as other observed phenomena. Therefore Einstein's theory has not been discarded. However, it is a complicated theory and the effects that it alone explains generally do not affect our daily lives.

Einstein's explanation of gravitation may, perhaps, have a certain philosophical appeal. But why is space curved in the presence of mass? The question "Why?" still crops up, and with each additional explanation the ultimate one just moves one step farther back.

PROBLEMS

1. Compute the gravitational attraction between
   a) the earth and the sun,
   b) the earth and a 70-kg man on the earth.

(See Appendix 2 for data.)

2. Compute the gravitational attraction in newtons between an electron,

$m_1 = 9.1 \times 10^{-31}$ kg,

and a proton,

$m_2 = 1.6 \times 10^{-27}$ kg,

when their separation is that in the normal hydrogen atom, namely 5×10^{-11} m.

3. The Cavendish experiment has recently been repeated using for the attracting masses a glass of water ($m_1 = 0.25$ kg) and a bag of sand ($m_2 = 12$ kg). Take the distance r between centers to be 25 cm. Compute

   a) the gravitational attraction in newtons and pounds of force,

---

* In curved space a beam of light, or a particle not acted on by a net force, follows a curved path rather than a straight line.

b) the resulting acceleration of the glass of water (which was suspended on the end of a meter stick supported by a long tape).

4. An airplane has a speed of 300 mi/hr or 440 ft/sec. The pilot wishes to turn as sharply as possible, but he cannot withstand an acceleration greater than $4g$, where g is the acceleration due to gravity.

a) Compute the radius of the smallest circle he can follow and the time it will take him to reverse his direction.

b) Repeat for a pilot flying 500 mi/hr.

5. Consider a pilot of mass m who wishes to fly vertical loop-the-loops.

a) Why must the centripetal acceleration at the top of a loop be at least equal to g?

b) If the inverted plane at the top of a loop experiences a centripetal acceleration of $3g$, does the pilot feel a force of $2mg$, $3mg$, or $4mg$ against the seat of his pants?

c) What force will the pilot feel if the plane takes the bottom of the loop with a centripetal acceleration of $3g$?

6. Show that for a banked curve in a highway there is one proper speed for a given angle of bank, and that this speed does not depend on how heavy the car is.

7. Call the earth's orbit 1 AU (astronomical unit). Use Kepler's third law to find the mean radius of the orbit of Venus and also that of Mars, given that Venus orbits the sun in 225 days and that Mars completes its orbit in 687 days.

8. Why would the creation of an artificial lunar satellite in an observable orbit be an accurate way of determining the moon's mass? (The moons of Jupiter have made it easy to compute Jupiter's mass.)

9. Compute the speed and period of a lunar satellite in an orbit whose radius is 6.7×10^6 m (a little over 4000 mi). Take the mass of the moon to be $M_E/81$ and refer to Section 4–8.

10. If the distance from the earth to the moon is $60 \times R_E$, where will a spaceship experience equal and opposite forces due to the earth and moon?

11. If a satellite is to serve as a space platform that hovers above a fixed point on the earth's equator, show that it must be placed in orbit about 22,000 mi above the earth's surface.

12. A satellite in orbit has a speed whose square is only half as large as it needs to be if the satellite is to escape from the earth's gravitational field. What speed would a satellite 186 mi above the earth (see Section 4–8) need to be given in order to make it escape?

References

COHEN, I. B.,   *The Birth of a New Physics*, Doubleday, 1960.

KOESTLER, A.,   *The Watershed, a Biography of Johannes Kepler*, Doubleday, 1960.

5

5|1 WORK The concept of work, like the concepts of length, time, and force, originated intuitively. Early civilizations were concerned with transportation and building. Men and animals were said to be working when they pushed and pulled objects against the forces of friction and gravity; in so doing they gradually became exhausted and had to be given food and rest. Since rest alone will not enable a laborer to keep going, it must be that the food he eats serves for him as fuel does for an engine.

Suppose that a certain engine does the job of lifting a 50-lb basket of bricks 20 ft. Say that the basket is lifted by a rope passing over a pulley to the engine (Fig. 5–1). The rope must exert a pull, call it F, on the basket.

The Conservation and Degradation of Energy Principles

If a second basket with the same load were connected to the same engine by means of another rope and pulley, the engine would exert a force F to lift the second 50 lb. If the two 50-lb loads were lifted by a single rope, the force needed would be $2F$. Should the engine lift the two baskets, we would say that it had done "twice the job" of raising one basket alone and this would obviously require the engine to use twice as much fuel as it did when only one basket was lifted. Thus the force needed and the fuel expended to do two equal jobs are here each twice what they are for one such job.

Should the engine lift a 50-lb basket first 20 ft and then another 20 ft we would say that it again did two equal jobs and must have used twice the fuel needed to do one such job.

Since we associate work done with the number of unit jobs accomplished and since the number of unit jobs accomplished is proportional to both the magnitude of the applied force F and the distance s its point of application moves (in the direction of the applied force), it is reasonable to define the work W that is done as

$$W = Fs. \qquad \qquad 5\text{-}1$$

This definition of work makes it not only proportional to the number of unit jobs done, but also proportional to the fuel expended in doing these jobs.

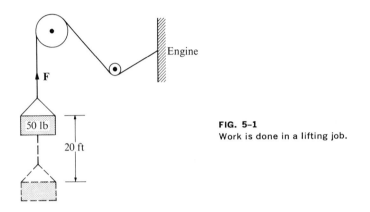

FIG. 5-1
Work is done in a lifting job.

In the mks system force is measured in newtons, distance in meters, and work in newton-meters. For brevity the word *joule* (abbreviated to J) has been substituted for newton-meter, that is,

$$1 \text{ joule} = 1 \text{ N-m}.$$

If the force $\mathbf{F}$ and displacement $\mathbf{s}$ are in different directions, the work W is defined as the product of the component, or projection, of the force in the direction of motion times the distance moved. Take the example shown in Fig. 5-2 in which the force $\mathbf{F}$ is applied diagonally downward in an attempt to push a trunk across the floor. Only the horizontal component $\mathbf{f}$ is effective and if the trunk moves a distance s, the work done is fs, not Fs. Suppose that the trunk is put on roller bearings; both the effective force needed and the work done will be reduced. Imagine that the bearings are greased until friction vanishes (a typical thought experiment); would not the work be zero in this case? It follows that no work is done

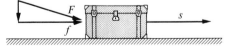

FIG. 5-2
Work done in pushing a trunk equals *fs*. (Although force *F* is exerted, only its horizontal component *f* is effective.)

while just holding a suitcase, or while moving it horizontally through the air as one's arm hangs vertically (Fig. 5–3).

FIG. 5–3
The vertical upward pull of the person's arm does no work as he moves horizontally.

Example. Suppose that the trunk in Fig. 5–2 has a mass of 10 kg, the applied force **F** has a magnitude of 5 N and the horizontal component **f** is 4 N. How much work is done if the trunk is pushed 8 m along the floor?

Solution. Since fs represents the work, we have

$$W = fs = 4 \text{ N} \times 8 \text{ m} = 32 \text{ J}.$$

This work is done by the agent applying the force regardless of whether the trunk is shoved at constant speed, is accelerated, or is allowed to slow down, although the result of the work is different in each case.

5|2 ENERGY

We have to do work to stretch a rubber band or elastic spring, because when we do so we pull the molecules farther apart. The molecules attract one another, so if we release our pull they move back closer together and the rubber band, or spring, contracts. In the contraction process the intermolecular forces, not we, do the work, work which may be used, say, to run a mechanical toy or lift a weight.

When fuel is burned the atoms in the molecules of fuel are rearranged, or they separate to recombine with other atoms. In so doing the intermolecular forces do a net amount of work which also may be used in some useful way. This is why an engine can be made to lift bricks and why the amount of fuel used is proportional to the work done by the system.

Let us say that when a system does the work W, it takes an amount of *energy* equal to W from some source such as fuel. According to this definition, energy represents the *capacity for doing work*. This agrees with our intuitive concept of energy, for when we are rested we say that we feel full of energy, or that we have more energy than we do after we have done a lot of work and become tired.

Of course, work is not always done by using the energy of some fuel. What about the case of a freely falling body? Gravity does work, but what is the source of energy for this work? Or, what is the source of energy for the work done against friction as a sliding object comes to rest? We shall see that we can answer these questions better after we consider the following question: What becomes of the energy used to do work? Our answer to this important question will be to say that there are many forms of energy besides that due to certain molecular arrangements found in fuels, that when a system A does an amount of work W, system A loses W units of energy of some form and the system B, on which the work is done, gains W units of energy, not necessarily in the same form. According to this view, work is a process through which energy is transported from one part of our world to another, and often transformed at the same time, but never created or destroyed. Figure 5–4 shows this schematically.

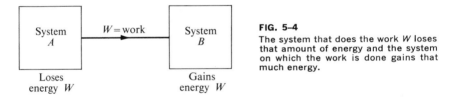

FIG. 5–4
The system that does the work W loses that amount of energy and the system on which the work is done gains that much energy.

This idea that energy can be moved around and transformed, but without changing the total amount in the world, is a postulate that must be tried and tested before we can say that it truly describes how our physical world behaves. If it holds up, then it may be relied upon and put to use. Conservation laws (statements that the value of something does not change when other things do) are particularly helpful when it comes to determining what events are possible and which ones are not. In the present case the problem has been one of recognizing the different forms of energy as they have been encountered. Considerable imagination and intuition have been called for, but the results have been successful. Let us look into how this has been done in a number of situations and hope that these examples will make the general principle clear.

5|3 POTENTIAL ENERGY

The earliest forms of energy recognized were those related to Newtonian mechanics. They are the energy associated with position, which we call *potential energy* (PE for short), and the energy associated with motion, which we call *kinetic energy* (KE for short). This section is concerned with the former.

Suppose that one wishes to lift a body of mass m slowly through a vertical height h near the earth's surface. We shall neglect air resistance. We shall also neglect the work done in getting the body in motion by assuming that the object is already moving up at constant speed, or that the speed is so low that the work in getting started is negligible. Then it is only necessary to apply an upward force just equal and opposite to the weight to keep the object moving (Fig. 5–5). Thus the applied force is mg up and it acts through the distance $s = h$; so according to Eq. (5–1) the work done is

$$W = Fs = mgh.$$

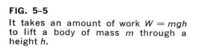

$W = $ work done by man $= mgh$

h

$F = mg$

FIG. 5–5
It takes an amount of work $W = mgh$ to lift a body of mass m through a height h.

Doing this work requires the expenditure of this much energy by the lifting agent, energy usually furnished by some kind of fuel. Thus the agent and his fuel supply, which we may call *system A*, lose the energy W. Where does it go? We are holding to the principle that it is not destroyed. The work performed is done on the object lifted, our *system B*, therefore we say that this object gained the energy W. Since the only thing about the body that changes is its height, we must assume that this energy is associated with the body's position in the earth's gravitational field. So we define the gain in gravitational *potential energy* as

gain in PE $= mgh$.   **5-2**

Two points should be understood in connection with this equation. First, it does not apply to rockets or satellites lifted so far above the earth that g is not constant; for such cases a different formula applies. Second, we have only defined the *gain* in PE of a lifted object, not the absolute value of its PE. Gain or loss in energy is all we have to account for in order to maintain the conservation of energy principle.

Example. A 50-kg person ascends (a) from sea level to the ground floor of a building 2 m above sea level, and (b) from the ground floor to a floor 3 m higher. Compute the gain in PE in (a) and (b) separately and the overall gain. (Fig. 5–6).

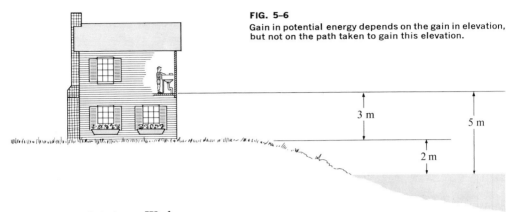

FIG. 5-6
Gain in potential energy depends on the gain in elevation, but not on the path taken to gain this elevation.

3 m

5 m

2 m

Solution. We have

$$mg = 50 \text{ kg} \times 9.8 \text{ m/sec}^2 = 490 \text{ N},$$

so that in (a)

Gain in PE $= 490 \text{ N} \times 2 \text{ m} = 980 \text{ J}.$

Similarly in (b)

Gain in PE $= 490 \text{ N} \times 3 \text{ m} = 1470 \text{ J}.$

In all, the person gained

$980 + 1470 = 2450 \text{ J}$ of PE

in going from sea level to the upper floor of the building, regardless of the path or time taken.

5|4 KINETIC ENERGY

Now, to eliminate the possibility of any change in PE, let us move a body horizontally. Since the object must be supported, we cannot avoid encountering some friction in actual practice and this is a complicating factor that is better left to later. Therefore, we return to the "thought experiment" of an object moving on a *frictionless* horizontal surface. This is a very good example of the use of the "thought experiment" because by eliminating friction from our considerations we can concentrate on the one thing we are now interested in, namely the relation between motion and energy.

Imagine an object of mass m at rest on a perfectly smooth horizontal floor. Let a horizontal force **F** be applied and push the body through the distance s (Fig. 5–7). Since **F** is the only force acting, it will start the body moving with an acceleration **a**, such that

$$\mathbf{F} = m\mathbf{a}.$$

In a time t the speed v that is gained and the distance s traveled from rest will be, respectively,

$$v = at, \qquad s = \tfrac{1}{2}at^2.$$

When we substitute v/a for t in the second equation we get

$$s = \tfrac{1}{2}a\left(\frac{v}{a}\right)^2 = \frac{1}{2}\frac{v^2}{a}.$$

The work done by **F** in the distance s is

$$W = Fs = (ma)\left(\frac{v^2}{2a}\right) = \tfrac{1}{2}mv^2.$$

Again we must recognize that the agent or system that applies the force **F** must lose the energy W and again we wish to postulate that such energy must go somewhere. This time the result of doing the work W is that an object has been set in motion, so we associate energy with this motion.

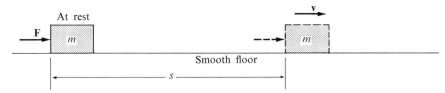

FIG. 5–7
Gain in kinetic energy occurs when an unbalanced force acts on an object and accelerates it.

We therefore define the *kinetic energy* gained from rest as equal to the work performed in the process, or

$$\mathrm{KE} = \tfrac{1}{2}mv^2. \qquad\qquad \textbf{5-3}$$

Here also two points should be noted. First, we have assumed that the mass m is constant and this is true at ordinary speeds, but not at ones approaching the speed of light (see Chapter 10). Second, we have defined *gain* in KE from rest, but since starting with no motion is starting with no KE, this gain in KE *is* the KE associated with the final motion.

Example. Let the force **F** in Fig. 5–7 be 4.5 N and the mass of the object be 0.5 kg. Find the KE and speed gained from rest in (a) $s = 2$ m, (b) $s = 8$ m.

Solution. The KE gained is the same as the work done by **F**, so that it is just Fs.

a) The KE gained is 4.5 N $\times$ 2 m $= 9$ J. From Eq. (5–3)

$$9 \text{ J} = \tfrac{1}{2} \times 0.5 \text{ kg} \times v^2,$$

$$v^2 = 36 \text{ m}^2/\text{sec}^2,$$

$$v = 6 \text{ m/sec}.$$

b) Here s is four times as great, and so are W and the KE gained, so that the KE $= 36$ J. Calling the speed gained v', we have

$$36 \text{ J} = \tfrac{1}{2} \times 0.5 \text{ kg} \times (v')^2,$$

$$(v')^2 = 144 \text{ m}^2/\text{sec}^2,$$

$$v' = 12 \text{ m/sec}.$$

Note that when the speed of an object is doubled its KE is quadrupled. Just as the applied force must act over four times the distance to produce twice as much speed, so must the braking force on a car act over four times the distance to bring a car to rest if it is going 60 mph rather than 30 mph. This is an important point in connection with motor vehicle safety.

5|5 INTERNAL ENERGY OR HEAT

Let us next eliminate the possibility of any change in either PE or KE. To do this we imagine that the horizontal force **F** is applied to an object of mass m that is moving on a *rough* horizontal surface. The force **F** is to be adjusted so that it just keeps the body moving at constant speed, like an engine pulling a train at a steady rate. At constant speed the acceleration, and hence the net force acting on the object, must be zero. This means that **F** is not the only force; there must be an equal opposing force, which we shall assume to be due to friction and call **f** (see Fig. 5–8). Since **f** and **F** are equal in magnitude,

$$f = F.$$

Suppose that the object moves a distance s in the direction in which the applied force **F** acts. Then again the system applying this force does the work $W = Fs$ and loses this amount of energy. Where does this energy go this time? The object did not move up or down, so its PE did not change. It did not go faster or slower, so there was no change in its KE. However, careful observation might disclose a slight warming of the object and the surface it traveled over. This heating effect may be noticed particularly when a rope slips rapidly through one's hand. We know that it takes energy to raise the temperature of a body, as when a pot of coffee is heated over a gas burner; this uses up fuel energy. Thus, if doing work against friction warms the surfaces that rub together, it must be that the energy lost by the system applying the force **F** has gone into what we call *thermal energy* or *heat*. The energy transformed into heat is $W = Fs$, or since $F = f$, we have that the

heat gained $= fs$,   **5-4**

where s is the distance moved against the frictional force f.

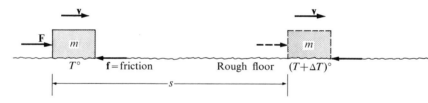

FIG. 5–8
Gain in thermal or internal energy occurs when work is done against friction.

What is heat? The term is usually understood to mean the energy associated with the motion and position of the individual molecules of a body. The molecules of any substance are supposed to be jumping around, rotating, and vibrating in a random way. The KE represented by such motion plus any PE due to intermolecular forces constitute what is called the *internal energy* (IE) of a body. This energy must be distinguished from the KE and PE of the body as a whole. For instance, when a block moves with a speed v, all of its molecules have this same average speed in addition to their random thermal motions. Internal energy is another name for thermal energy, and heat is the more colloquial term for both.

Equation (5–2) expresses gain in PE in terms of the change in height h, while Eq. (5–3) gives the gain in KE in terms of the speed attained. Can we express gain in internal energy in terms of similar parameters

111

associated with an object? Experimental observation indicates that when we give a body heat we affect it in one of two ways: either (a) we raise its temperature T,* or (b) we change its state, as when ice melts or water boils. In both processes the amount of heat needed is naturally proportional to the mass m of a given substance to be altered in a given way. In (a) the heat required per unit mass for a given temperature rise is quite different for various substances, but for a given substance, like copper, it is very nearly proportional to the rise in temperature ΔT.† We may then write, *for a change in temperature,*

gain in internal energy $= cm \, \Delta T$, **5-5**

where c is approximately constant for a particular substance. We call c the *specific heat capacity*, or just the *specific heat*, of the material. Values of c for various materials have been determined experimentally and listed in tables found in handbooks and many texts. For instance, by stirring water with paddles one may measure the amount of work needed to raise 1 gm of water 1°C‡ (Fig. 5–9); it is found to be about 4.19 J. Since water is such a common substance in our world and in our lives, 4.19 J was given its own name and called a *calorie* (or cal). One thousand calories

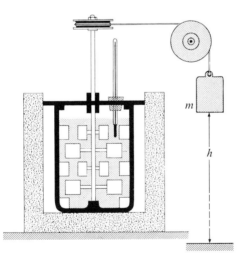

FIG. 5–9
By letting m fall N times the work $Nmgh$ is done. This work is used to rotate a paddlewheel in a certain mass of water and the resulting rise in temperature is measured.

* While we all have an intuitive conception of temperature in terms of "hot" and "cold," a more physical definition is given in Section 5–11.
† Temperature scales are defined in Experiment 9 at the end of the chapter.
‡ °C means degrees centigrade (or Celsius).

will heat 1 kg of water 1°C, and this *kilocalorie* (kcal) is also the unit of energy referred to in cookbooks and in connection with dieting.

With regard to (b), change in state, the heat required is just proportional to the mass m, but the constant again varies from one material to another. We have, *for a change in state,*

$$\text{gain in internal energy} = lm, \qquad\qquad \textbf{5-6}$$

where l is called the *latent heat* of the substance. Eighty calories will melt 1 gm of ice and at 100°C, 539 cal will boil 1 gm of water. (Such specific numbers can always be found in tables and should be memorized only when one is going to use them frequently.)

Example. A 500-gm block of copper, for which $c = 0.093$ cal/gm-°C, is pushed 10 m at constant speed along a horizontal floor by a horizontal force of 6 N. Find (a) the heat developed and (b) the rise in temperature of the copper if it gains two thirds of this heat.

Solution

a) We have $F = f = 6$ N. From Eq. (5–4) the heat developed is $fs = 6$ N $\times$ 10 m $= 60$ J, which is

$$\frac{60 \text{ J}}{4.19 \text{ J/cal}} = 14.3 \text{ cal.}$$

b) The copper gains $\frac{2}{3} \times 14.3$ cal $= 9.5$ cal of internal energy, which results in a rise in its temperature. From Eq. (5–5),

$$9.5 \text{ cal} = 0.093 \frac{\text{cal}}{\text{gm-°C}} \times 500 \text{ gm} \times \Delta T,$$

$$\Delta T = 0.2\text{°C.}$$

Evidently the block would have to be moved back and forth several times before it would be noticeably warmer.

5|6 THE CONSERVATION OF MECHANICAL ENERGY IN FRICTIONLESS SYSTEMS

Earlier we asked the question: What is the source of the KE that a falling body gains? The answer is that the gain in KE is at the expense of an equal loss in the PE of the body.

Consider an isolated system, that is, one which does not exchange energy with the rest of the world, as, for example, a frictionless pendulum

(Fig. 5–10). Such a pendulum would swing back and forth indefinitely. At the top of its swing its KE is zero because it comes to rest for an instant. At the bottom of the arc it has maximum speed and KE, but it has lost height and PE. If we postulate that the gain in KE equals the loss in PE, we are really saying that the sum of these two kinds of energy remains constant, or

KE + PE = a constant.

This theorem is called the *conservation of mechanical energy*. This must hold for an isolated and frictionless system if we are to maintain the more general principle that energy may be transformed, but cannot be created or destroyed.

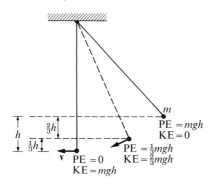

FIG. 5–10
The conservation of mechanical energy. The KE gained equals the PE lost. (The PE has arbitrarily been taken to be zero at the bottom.)

Example. An object of mass m falls from rest a distance h. What speed does it attain if air resistance is negligible?

Solution. Since mgh is the PE gained when the object is lifted through the height h, this will also be the PE energy lost in the fall. If v is the speed attained, the KE gained is $\frac{1}{2}mv^2$. Equating gain in KE to loss in PE gives the equation

$$\frac{1}{2}mv^2 = mgh.$$

The factor m cancels out and we get

$$v = \sqrt{2gh}. \qquad \textbf{5-7}$$

The mass, or weight, of the object makes no difference when air resistance is negligible.

5|7 THE CONSERVATION OF ENERGY IN MECHANICAL PROBLEMS INVOLVING FRICTION

We saw that when the point of application of an applied force **F** moves a distance s against a frictional force **f**, an amount of work, $W = fs$, is done against friction and this represents a gain in the internal energy of the bodies in contact, at the expense of the energy of the agent applying the force **F**. But work may be done against friction *without* the action of an applied force. So long as there is a displacement s against a frictional force **f** there will be a gain of fs units of internal energy or heat. Without an outside agent this energy must be developed at the expense of some other form of energy within the system, if the conservation of energy principle is to be maintained. So we say that the heat developed as an object slides steadily down a rough slope comes from a corresponding loss in the object's PE, and that the heat developed when brakes are applied to slow a moving car comes from the loss in KE of the car.

To sum up, let us take the general case in which an outside agent causes a force **F** to act on an object, of mass m, through a distance s, with the possibility that work is done lifting the object, speeding it up, and overcoming friction. The conservation of energy principle states for this case that the

energy supplied by agent = gain in (PE + KE + IE) of object.

Stated mathematically, this becomes

$$Fs = mgh + (\tfrac{1}{2}mv_2^2 - \tfrac{1}{2}mv_1^2) + fs, \qquad\qquad \textbf{5-8}$$

where v_1 is the initial speed and v_2 the final speed. From this general case one may proceed to many simpler ones, such as one for which $\mathbf{F} = 0$ (no outside agent), or $h = 0$ (horizontal motion), or $v_1 = 0$ (start from rest), or $\mathbf{f} = 0$ (no friction).

Example 1. An applied force $\mathbf{F} = 20\,\text{N}$ parallel to an inclined plane pulls a 1-kg object 2 m up the plane (Fig. 5–11). If the object rises 1 m and its speed increases from 3 m/sec to 7 m/sec, what was the force of friction?

Solution. The energy supplied by the outside agent is

$$Fs = 20\,\text{N} \times 2\,\text{m} = 40\,\text{J}.$$

$$\text{Gain in PE} = mgh = 1\,\text{kg} \times 9.8\,\frac{\text{m}}{\text{sec}^2} \times 1\,\text{m} = 9.8\,\text{J}.$$

$$\text{Gain in KE} = \tfrac{1}{2}mv_2^2 - \tfrac{1}{2}mv_1^2 = \tfrac{1}{2}(7^2 - 3^2)\,\text{J} = 20\,\text{J}.$$

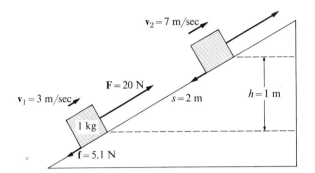

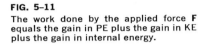

FIG. 5–11
The work done by the applied force **F**
equals the gain in PE plus the gain in KE
plus the gain in internal energy.

The gains in PE and KE account for 29.8 J of the 40 J supplied. The remaining 10.2 J must have gone into heat, as represented by fs. Therefore

$$f \times 2 \text{ m} = 10.2 \text{ J}, \qquad f = 5.1 \text{ N}.$$

Example 2. *Machines.* A machine is a device whereby an applied force, or "effort," **F** moves a distance s and does something like raising a weight mg through a height h. Refer to Eq. (5–8) and assume no change in KE. Then we see that mgh, the useful work accomplished, cannot exceed Fs, the work put in, because the term fs is always positive. However, by making h less than s, a small effort **F** may lift a large weight mg. The automobile jack illustrates this point; see also Fig. 5–12.

FIG. 5–12
While the weight lifted by a machine
may greatly exceed the effort applied,
the useful work accomplished is
actually less than the work put in.

Weight lifted

5|8 HEAT FLOW

We have seen that energy may be transferred through work. Is this the only way? How is energy transferred from the sun to the earth, or from the hot end of a metal rod to the cold end? We must recognize a second method of energy transfer based on a temperature difference and involving a transfer of internal energy, i.e., *heat flow*. Actually this may occur in three ways, namely through *conduction* (atom by atom), *convection* (flow of a fluid from a hotter to a colder region), and *radiation* (electromagnetic waves moving through space with the speed of light).

5|9 THE CONSERVATION OF ENERGY PRINCIPLE. FUNDAMENTAL LAW IV

What we have postulated so far in this chapter may now be summed up as follows:

Through (a) *work, or* (b) *heat flow, energy may be transferred and transformed, but it cannot be created or destroyed.* In other words, *the total energy content of a closed system is constant.*

This is our *Fundamental Law IV*, and it embodies the so-called *First Law of Thermodynamics.* Note that work involves a force acting through a distance and that heat flow requires a temperature difference.

There is an interesting analogy between energy and money. We have seen that there are various forms of energy just as there are different currencies. Energy may be changed from one form to another, while dollars can be changed into French francs, German marks, etc. In currency exchange the ratio may or may not be firm, that is, the same everywhere and from one day to the next. With firm rates one may change dollars into francs, the francs into German marks, the marks back into dollars, but if commissions are either neglected or their sum included, one will end up with the original number of dollars. There was a time, however (around 1950), when one could buy on the black market 500 old francs for one dollar, while a bank would give one back his dollar for 350 old francs. Thus $7 could be changed into 3500 francs via the black market and turned back into $10 via normal channels. This is termed *arbitrage* in financial circles. Now the conservation of energy principle states that there is no arbitrage when energy is changed from one form to another. One always gets a joule for a joule. Of course there may be some dissipation of energy and this corresponds to the commission in financial exchange, but such energy is not lost. Energy *inflation* is also ruled out, for although printing presses may create new money, no process has been found for creating energy.

Reference has been made to the fact that new forms of energy have been invented from time to time to explain what would otherwise be an unaccountable appearance or disappearance of energy. Thus potential energy, kinetic energy, internal energy, fuel energy, etc. have been introduced. Let us consider one of the latest forms of energy to be postulated.

Illustration. *The neutrino*

It has been found in connection with radioactive beta decay that a nucleus *A* may decay into nucleus *B* and an electron (Fig. 5–13a), with the energy *E* released in the decay appearing as KE of the electron. On the other hand (Fig. 5–13b), a nucleus identical with *A* may decay into one just like *B* and an electron, the electron receiving a KE of *E*/2. In a third case, the KE of the electron may be only *E*/4. Where does the missing energy go?

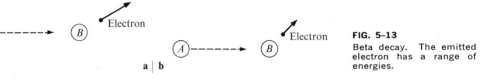

FIG. 5–13
Beta decay. The emitted electron has a range of energies.

a | b

Physicists could have said (and some did) that here energy was just destroyed and that this was a case in which the conservation of energy principle failed. However, Fermi proposed that the missing energy was not lost, but was carried away by an elusive particle first suggested by Pauli and called the *neutrino* (see Fig. 5–14). A neutrino is supposed to have no mass or charge, so that it can slip away unobserved. While at first thought this proposal may seem amusing, be assured that it is taken very seriously by most physicists. This is because (1) the neutrino concept has simplified our understanding of many nuclear phenomena; (2) it has led to several successful predictions, and finally (3) more than 20 years after the neutrino's existence was postulated the first direct experimental detection of neutrinos was obtained by observing the rare capture of neutrinos by matter.

FIG. 5–14
Neutrino emission in beta decay.

5|10 THE UNAVAILABILITY OF ENERGY. FUNDAMENTAL LAW V

Energy is stored in coal and oil, and can be tapped when we want to use it to do work. If this work is done against gravitational or elastic forces, the original fuel energy is transformed into energy associated with the position or stretch of a body, but it is still stored in an available form and so we call it potential energy. Even kinetic energy may be used to do work, as when wind propels a windmill or sailboat, or it may be turned into stored potential energy, as when a pendulum bob swings back up. However, when work is done against frictional forces and heat is developed, the heat energy tends to spread out and be dissipated. This dissipation is the result of heat flow. When surfaces rub together they become warmer than their surroundings and heat always flows from a warmer to a colder region. Insulation may retard the flow, but it cannot reverse it. This is an empirical fact. The warmer surfaces cool, the surroundings warm up, and they approach a common final temperature. When work is done against frictional forces, a system will possess more internal thermal energy than originally, but usually it can do nothing with this energy. It has become unavailable for doing further work.

Mention has been made of the experimental fact that heat naturally flows from a warmer to a cooler body. The reverse process may be made to occur, but only if some agent does work and uses available energy. A refrigerator, for example, moves heat energy from the cold box to the warmer room, but it takes electricity or gas to run it. In the natural process of heat flow from a hot to a cold region useful work may be done for us, that is, *some* of the heat taken from the hot source may be turned into useful work and only the remainder wasted. This is the principle of the heat engine. We know from experience that in all heat engines the wasted energy is thrown out in the exhaust or given to some cooling system that carries it away.

Since frictional forces can never be completely eliminated, work cannot be done without some energy being dissipated. Heat flows of its own accord from a hotter to a colder body and this results in dissipation of useful energy, and hence heat flow also reduces the amount of available energy in our world. We thus arrive at the following general principle:

As a result of natural processes, the energy in our world available for work is continually decreasing.

This is called the *degradation of energy principle*, also the *second law of thermodynamics*, and it constitutes our *Fundamental Law V*.

5|11 THE STATISTICAL INTERPRETATION OF THE DEGRADATION OF ENERGY

One of the great theories of physics is the kinetic theory of gases. In this theory one starts by making some educated guesses or restricted hypotheses about gas molecules, notably that gas molecules are in continual random motion and that they collide as do hard elastic spheres. The *pressure of* a gas is attributed to the impacts of the molecules against the walls of the containing vessel; at each impact a molecule reverses its momentum toward the wall to momentum away from the wall and thereby exerts a force against the wall. The *temperature* of a gas is taken to be a measure of the mean thermal energy of the molecules. *Absolute zero* is defined as the temperature at which molecular motion would cease, so that at absolute zero the molecules would exert no pressure even if they were still free to behave as in a gas. On the absolute temperature scale proposed by Lord Kelvin, temperature is measured up from absolute zero, being proportional to the pressure of a gas whose volume is held constant. The Kelvin temperature T is then also proportional to the mean KE of translation per molecule at that temperature; since the constant of proportionality is the same for all molecules, if we write

$$\text{Mean KE of translation per molecule} = \tfrac{3}{2}kT, \qquad \textbf{5-9}$$

we see that k, which is called *Boltzmann's constant*, must be a universal constant. (The factor $\tfrac{3}{2}$ need not concern us.)

The formulas derived from kinetic theory have been tested experimentally and found to hold well provided the gas is not compressed into too small a volume. At close distances the molecules are found to exert forces on one another and their finite size must be considered. The best experimental value for k is

$$k = 1.38 \times 10^{-23} \text{ J/molecule-}°\text{C}.$$

In liquids and solids the thermal energy takes forms other than KE of translation of the molecules. In a solid the molecules or ions vibrate back and forth, but this energy also increases with the temperature.

While the temperature determines the *mean* thermal energy of a molecule, as in Eq. (5-9), some molecules will have more and some less energy than the average. This results from the random motion of the molecules. When two molecules collide, KE may be passed from one to another, as in a collision between billiard balls. However, if the temperature of, say, a gas is uniform throughout, the fast, the medium-speed, and the slow molecules are pretty uniformly mixed throughout, as are the cards of one suit in a well-shuffled deck.

Now consider, as an example of a natural process involving the degradation of energy, the situation in which a warm and a cold body come in contact. The molecules of the warm body are, on the average, moving faster than those of the cold body (Fig. 5-15). We start with an ordered arrangement in which the fast and slow molecules have been sorted out, as it were. Upon contact the faster molecules and slower molecules make impacts on one another and on the average the fast molecules slow down and the slow ones speed up until the mean speeds of the molecules in each body are the same. Thus the two bodies come to a common final temperature, which will represent a state of equilibrium.

FIG. 5-15
Here the fast and slow molecules have been sorted out, but they will not stay that way if left to themselves.

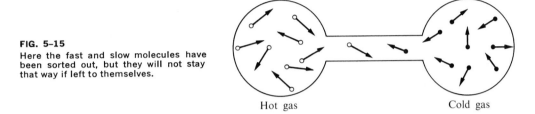

Hot gas   Cold gas

As another example, take the case of air rushing out of a tire. Since the pressure and density outside the opened valve are less than inside, fewer molecules will happen to be heading into the tire than are heading out. When equilibrium is reached, the numbers heading in and out will be the same because the pressure and density of the air on each side of the open valve will be the same.

In the above examples it is easy to explain statistically why the processes tend to proceed in only one direction. In a closed room the molecules of air do have different speeds, but the molecules moving faster than average are not likely to be found in one part of the room and the slower molecules in another part. Why? Because such a distribution is *improbable*, which means that according to the laws of chance, there are far fewer random distributions that are equivalent to a sorting out of fast and slow molecules than there are random distributions that are equivalent to a nearly complete mixing of fast and slow molecules. For the same reason the density of molecules in a room is not likely to fluctuate much from the mean density, so that the local pressure (if we could measure it) is not likely to vary much from the observed overall mean pressure. Thus a tire is unlikely to inflate itself.

When dealing with a large number of individual objects, arrangements that involve little order can be obtained in many more ways than specific ordered arrangements. Hence if molecular motions are subject to the laws of chance, then disordered arrangements are much more likely to occur than ones in which the fast molecules are sorted out from the slow ones, etc. This leads to the following statement of the second law of thermodynamics.

As a result of natural processes, the order in our physical world is continually decreasing.

In a sense our world is "running down." Physicists and chemists have introduced a mathematical function called *entropy* to measure how much the energy of a system has been degraded. The higher the entropy the more energy *un*available for work will be found in the system. We are using up available chemical energy in our fossil fuels and available nuclear energy in our uranium and hydrogen. We shall never get this available energy back. On the other hand, the sun continually furnishes us with more available energy in the form of direct sunlight, the energy of wind and rain, and the chemical energy of growing plants. This is possible because the sun's temperature is much higher than that of the earth. Eventually (after billions of years) the sun will burn itself out and cool down. Suppose that in the meantime all of the earth's sources of PE are depleted. Then, even if the sun gave up all its available energy and thereby warmed our world throughout to a comfortable 70° Fahrenheit (F), life could not exist, since there would be no *difference* in temperature anywhere and hence no possibility of doing useful work.

The British astronomer Hoyle and others have suggested that there may be distant regions of the universe in which the second law of thermodynamics does not hold true; in such regions available energy could be built up, as may once have been the case for our world. After all, the second law of thermodynamics only forbids events that are *improbable*, not ones that are *impossible*.

Can an intelligent being violate the second law of thermodynamics? Most assuredly! We do so when we arrange cards in a deck, put bricks together to form houses, set up corporations, etc. The British physicist, James C. Maxwell, imagined a demon who, placed at the open valve of a tire, would shut the valve when a molecule approached him from within the tire, but would open the valve to molecules entering from outside and thereby inflate the tire without pumping. The second law rules out the existence of such demons in the physical world. A Supreme Being could certainly violate the second law, but would He have to, or want to, in order to run the world wisely? He would not, it may be contended, because when a sufficiently large number of individual particles and events are involved, then the laws of chance become most reliable laws. When fluctuations about the most probable situation are percentagewise negligible, one may take the occurrence of this most probable situation as a practical certainty. This being the case for thermal processes in our physical world, what difference does it make whether they are governed

122

by fixed laws or by the laws of chance? In many instances reliance on the laws of chance proves to be the most practical procedure. As Alexander Pope put it in *An Essay on Man:*

Remember, man, the Universal Cause
Acts not by partial, but by gen'ral laws;
And makes what happiness we justly call,
Subsist, not in the good of one, but all.

Illustration 1

A government passes a law that applies to all of its citizens. The law may be unfair to a few individuals, but prove to be beneficial to the vast majority. Substitution of a set of laws, one for each individual citizen, would impose a costly burden of administration and be quite impractical.

Illustration 2

In a new deck of playing cards the cards are all sorted by suit and number. In the course of shuffling, the original order is *very* unlikely to return again. However, any arrangement obtained is no more likely than the original one! What, then, does shuffling accomplish?

If a shuffled deck is dealt to four hands in a game of bridge, there are certain *classes* of hands that can occur in many more ways than others. Thus you are much more likely to obtain a hand headed by a 5-card suit than one headed by a 9-card suit. There are enough cards in a bridge deck so that the laws of statistics apply well if the shuffling is thorough; the result is a definite type of game, one that has proven immensely popular for many years.

PROBLEMS

1. Compute the work done against gravity when a 75-kg man climbs

   a) stairs rising 4 m, and

   b) a mountain whose summit is 400 m above its base.

2. The man in problem 1 could easily climb the stairs in 8 sec, while 70 min should be allowed for the mountain climb. Find the rate at which the man does work (this is his power output) in each case. Express your answers in J/sec (or watts) and convert to horsepower (1 HP is about 750 J/sec).

3. How much work is needed to push an object of 10-kg mass at constant speed up a frictionless inclined plane that is 1 m long and 0.5 m high? Does it matter whether the applied force is up the plane or horizontal?

4. How much work is needed to push the 10-kg mass at constant speed up an inclined plane 1 m long and 0.5 m high if there is a frictional force of 10 N? Account for what becomes of the energy transferred by this work.

5. A horizontal force of 50 N is applied to an object whose mass is 5 kg. The force of friction is 10 N.

    a) What is the work done against friction if the object moves horizontally 10 m.

    b) Compute the gain in kinetic energy.

6. Show that when an object moves horizontally under an applied force **F** and an opposing frictional force **f**, then the gain in KE equals the product of the magnitude of the net force, $F - f$, and the distance moved.

7. Show that Eq. (5–7) agrees with the kinematic expressions for a freely falling body, that is,

$$v = gt \quad \text{and} \quad s = \tfrac{1}{2}gt^2.$$

8. Assume that the brakes of a car exert a constant retarding force.

    a) Find the ratio of the distance needed to stop a car going 75 mi/hr to the distance needed to stop the car when it is going 30 mi/hr.

    b) Find the ratio of the corresponding times in (a).

    c) For a given braking force, how does the distance depend on the weight of the car?

9. A gun of mass M fires a bullet of mass m; Q units of chemical energy are transformed into KE. What fraction of Q does the bullet receive? [*Hint:* Remember the conservation of momentum theorem!]

10. An elastic collision is one in which no KE is lost. When freight cars couple together, is the collision elastic? Prove your answer.

11. A 500-gm piece of copper is dropped 6 m into a pail of sand. Assume that 10% of the heat generated goes into the copper and compute its rise in temperature.

12. In a heat engine the gas in the cylinder comes back to its original state after each *cycle*. Suppose that in one cycle the gas takes in 800 J from a heat source and performs a net amount of work equal to 200 J.

    a) How much heat must the gas throw out per cycle?

    b) Since the 800 J must be paid for, what percent of the energy paid for goes into work?

13. Devices that change energy from one form to another are called *transducers*. Name a common transducer that will change

   a) sound energy into electrical energy,

   b) electrical energy into sound,

   c) light energy into electrical energy,

   d) chemical energy into electrical energy,

   e) electrical energy into mechanical energy, and

   f) internal thermal energy into mechanical energy?

14. Suppose that a machine gun fires forty 10-gm bullets per second at a target whose area is 0.5 m^2 and that the initial speed of a bullet is 500 m/sec.

   a) Express all quantities in mks units and find the pressure (force per unit area) exerted on the target if the bullets all become embedded in the target.

   b) Compare this with the pressure due to molecular impact on the walls of a vessel containing a gas at a pressure of

   1 atmosphere (atm) $= 1.01 \times 10^5 \text{ N/m}^2$.

15. Find the ratio of the chance of getting 50% "heads" to that of getting 75% "heads" when

   a) a coin is tossed four times, and

   b) repeat for the case where the coin is tossed eight times.

What conclusion can you draw from your calculations?

INCLINED PLANE
The Conservation of Energy

Experiment 8

Object: To verify the principle of conservation of energy for a body being pulled up an inclined plane.

Theory: Consider a block of mass m on an inclined plane; a string attached to the block runs over a pulley at the top of the plane and supports a hanging weight. If the plane were perfectly smooth or frictionless, the block would be in equilibrium if the string exerted a pull up the plane equal and opposite to the force exerted down the plane by the weight of the block. Call this force $\mathbf{F}_0$.

Next consider the actual, somewhat rough plane. To pull the block at constant velocity up the plane (Fig. 5–16a) will require a force

$$F_u = F_0 + f, \tag{1}$$

where f is the frictional force. Letting the block slide down the plane at constant velocity (Fig. 5–16b) will require a force up the plane equal to

$$F_d = F_0 - f, \tag{2}$$

since f always acts opposite to the motion and so acts up the plane when the block slides down it. Then

$$F_u - F_d = 2f, \tag{3}$$

so that by measuring F_u and F_d one may compute f.

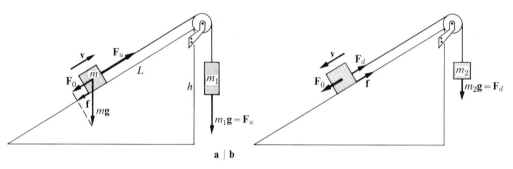

FIG. 5–16
a| Block sliding up an inclined plane: $\mathbf{F}_u = \mathbf{F}_0 + \mathbf{f}$.
b| Block sliding down an inclined plane: $\mathbf{F}_d + \mathbf{f} = \mathbf{F}_0$.

Now suppose that the force F_u is used to pull the block at constant speed the whole length L up the plane. The work done is

$$W = F_u L. \tag{4}$$

Part of this is expended in increasing the potential energy of the block, while the rest is done against friction. (At constant speed there is no change in KE.) The work done against friction is $W_f = fL$, and we assume that this work is completely converted into thermal energy. The potential energy of the block is increased by

126 $$PE = mgh, \tag{5}$$

where h is the vertical rise or height. Thus if energy is conserved, we have

$$W = \text{PE} + W_f, \tag{6}$$

$$F_u L = mgh + fL. \tag{7}$$

Our problem is to test Eq. (7) experimentally.

Procedure

Step 1. Determine m_1 and m_2 experimentally (see Fig. 5–16); these are the masses of the hanging weights that allow slow motion up and down the plane, respectively. Remember that the motion of the block must *not* be accelerated or there will also be a gain or loss of kinetic energy. It is not necessary to have the block slide the whole length L of the plane, for if it slides a distance of, say $L/2$, then the vertical rise will be $h/2$; Eq. (6) becomes

$$\frac{F_u L}{2} = mg\frac{h}{2} + f\frac{L}{2},$$

which is equivalent to Eq. (7). So it is best to choose a section of the plane for which friction is as nearly constant as possible and to determine both F_u and F_d over this same section. Once the F's have been obtained one may "in thought" suppose that the block travels the whole length up the plane and rises the full height, because it is easier to measure L and h than the actual length and height used. So measure L and h.

Step 2. Calculate F_u and F_d in newtons. Use Eq. (3) to compute f in newtons.

Step 3. Compute the terms $F_u L$ and fL in joules. Weigh the block to obtain its mass in kg. Calculate the term mgL.

Step 4. Check Eq. (7), noting the numerical difference between the value of one side and that of the other.

Step 5. Try to justify this difference by considering the possible error in each term of Eq. (7). You will have to assign possible errors, based on experimental observation, to F_u and F_d. What quantities are mainly responsible for error: F_u, F_d, L, h, or m?

THE GAS THERMOMETER
Experiment 9
Temperature Scales

Object: To calibrate a gas thermometer in degrees Kelvin (°K) and degrees centigrade (°C).

Theory: The ideal temperature scale should not depend on the thermal properties of any particular substance. The change in pressure of a fixed volume of a dry gas, such as air, furnishes an easy practical means of measuring temperature on a scale that closely approximates the ideal.

In defining a temperature scale one must make two further arbitrary choices: (1) fix the size of the interval which is to be called one degree, and (2) choose the zero point of the scale. For the Fahrenheit scale we choose (1) one degree equals $\frac{1}{180}$ of the interval between the freezing and boiling points of water, and (2) zero is 32 degrees below the freezing point of water. For the centigrade (also called Celsius) scale we choose (1) one degree equals $\frac{1}{100}$ of the interval between freezing and boiling, and (2) zero is the freezing point of water. For the Kelvin scale we choose (1) one degree equals $\frac{1}{100}$ of the interval between freezing and boiling, (2) zero is the temperature at which molecules would cease to move and so would exert no pressure. Since the last represents a sort of ultimate in coldness, it is called *absolute zero*. Degrees Kelvin are also called *degrees absolute*.

We shall calibrate a constant volume gas thermometer and use it to measure room temperature and to compute absolute zero in degrees C.

Procedure

Step 1. Your gas thermometer bulb has been in the room since the previous day. The mercury pressure gauge attached will thus be ready to give readings corresponding to room temperature; hence note both mercury levels before you touch the bulb in any way. (These levels do not have to be the same.)

Step 2. Surround the bulb with finely crushed ice and then flood the ice with ice water. Bring the mercury in the closed tube (next to the bulb) to where it was in Step 1—this keeps the gas at constant volume—and then read the level in the open tube.

Step 3. Surround the bulb with hot water and heat to boiling. Again bring the mercury in the closed tube to the same level as before and then read the level in the open tube.

Step 4. Read the mercury barometer. Compute in centimeters of mercury the absolute pressure of the gas in the bulb at freezing, at boiling, and at room temperature. Consider carefully in each case whether to add the difference between the open and closed tube readings to the barometer reading, or whether to subtract the difference from the barometer reading.

Step 5. Construct a graph of pressure p vertically against temperature T horizontally. Let the boiling point be at the extreme right and plot its gas bulb pressure first. Then move 10 squares to the left, call this the freezing point, and plot its gas bulb pressure. Connect these two points with a straight line and extend the line down to where it crosses the p equals zero, or temperature, axis.

Step 6.   Label the steam point 100°C and the ice point 0°C, then compute (by counting squares) absolute zero in degrees C.

Step 7.   Call the zero pressure point zero degrees Kelvin (0°K) and compute the freezing and boiling points in °K, or absolute degrees.

Step 8.   Use your graph, or a proportionality, to compute the room temperature from the pressure of the gas bulb at room temperature. That is, treat room temperature as some unknown temperature you wish to determine.

Step 9.   If desirable, a correction may be determined and made for the fact that the air in the closed tube between bulb and mercury was not cooled or heated and so "diluted" the pressure difference between freezing and boiling by possibly one or two percent.

6|1 DEFINITION OF A WAVE

A change in the state or in the condition of a medium may be called a *disturbance* of the medium. Then we may say that *a wave is a disturbance that is propagated through space.* Since there are many ways of disturbing a medium, there are many kinds of waves, and a description of the physical world would not be complete without summarizing the properties of waves in general. Waves are important in the theories of sound, light, electricity, and magnetism, and even atomic physics.

After classifying waves to some extent, we shall turn to the study of wave propagation, about which many experimental facts are known. Our problem will be to show how all of these facts may be summarized as theorems derived from a single basic principle or postulate.

Huygens' Principle of Wave Propagation

The disturbance in a wave may be an actual displacement of a material medium, as in the case of water waves, sound waves, and waves in vibrating strings, or the disturbance may be simply a change in a field (electric, magnetic, or possibly gravitational) in empty space, as in the case of light and radio waves. However, in all waves, something is passed on; there is a connection between the disturbance at one point in space and the disturbance at a neighboring point. This connection may be due to intermolecular forces and so be related to the elastic properties of a material medium or, as in the case of waves in empty space, it may be a property of the field that is being propagated. For wave propagation a changing displacement or varying field at a point A must result in a similar sort of disturbance at a neighboring point B.

Waves are never static. A diagram of a wave shows only what the disturbance is at various points in space at one given instant of time. A picture of a wave is like one frame in a moving-picture reel; the next frame will show a little different picture, and so on. Figure 6–1 is such a picture; it shows a pulse of transverse displacement in a rope as viewed at a particular instant. This is called a *waveform* diagram; in it the disturbance at a given instant is plotted as a function of position in space. But this does not show the motion of the disturbance. Is it moving toward or away from P? How fast is it traveling? To describe a wave more completely one may show waveform diagrams for different times, i.e., show a moving picture of the wave.

Time t_0

P

x

FIG. 6–1
Waveform diagram of a transverse pulse in a rope.

It is often possible to describe a wave most concisely in terms of mathematics, particularly if the wave is periodic, or repeats itself. Such a description is given by a *wave function*, which expresses the functional relationship between the disturbance, which we shall call Ψ, the position in space (such as the distance x measured along a rope), and the time t.

6|2 TYPES OF WAVES

There are many ways of classifying waves; the following are offered in order to point out in what ways waves may differ.

a | Mechanical versus field waves

Mechanical waves include all kinds in which material particles are displaced. Sound waves, earthquake waves, water waves, and waves in strings all involve the motion of the particles of the medium through which they pass.

Field waves include those in which electric and magnetic fields are propagated through space. Our most familiar examples are the various kinds of electromagnetic waves (radio, infrared, visible light, ultraviolet, x-rays, and gamma rays). Such waves may pass through material media, but they do not depend on the presence of matter for their propagation and are transmitted best through a vacuum. Such waves all have the same speed $c = 3 \times 10^8$ m/sec (or 186,000 mi/sec) *in vacuo*.

b | Transverse versus longitudinal waves

A *transverse* wave is one in which the disturbance is directed at right angles to the direction in which the wave is being propagated (see Fig. 6–2). A taut rope may be shaken in any plane that passes through the rope. In a vertical rope the disturbance may be east and west, north and south,

132

or a combination of these. If the vibration occurs only in one plane, as it would have to if a rope were passed through a picket fence and shaken, the wave is said to be *polarized* in that plane. Waves in strings and ripple waves on the surface of a liquid are familiar examples of transverse waves. The polarization of light indicates that light is propagated as a transverse wave.

FIG. 6-2
Transverse waves in a rope.

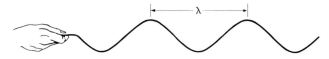

In a *longitudinal wave* the disturbance occurs parallel to the direction in which the wave is proceeding. Individual particles move back and forth, producing alternate regions of compression and rarefaction. Such waves cannot be polarized. Sound waves are our most familiar example of longitudinal waves. (See also Fig. 6–3.)

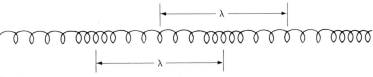

FIG. 6–3
Longitudinal waves in a coil spring.

There are also types of waves in which the disturbance is neither transverse nor longitudinal. For example, in deep-sea water waves, the particles move approximately in circles (Fig. 6–4); such motion may be viewed as a combination of longitudinal and transverse motion.

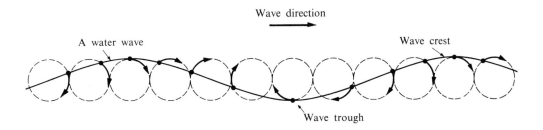

FIG. 6–4. Water molecules move in circular orbits about their original positions when a wave passes by. At the crest of a wave the molecules are moving in the direction the wave is traveling, while in the trough the molecules are moving in the opposite direction. There is no net motion of water involved in the motion of a wave.

133

c | Pulses versus sustained waves

Figure 6–1 shows a *pulse*, or disturbance of short duration. Pulses include shock waves, such as those set up in air by jet aircraft flying at speeds greater than that of sound. Shock waves are also produced in air by explosions, in water by speedboats, in rocks by earthquakes and blasting. A line of falling dominoes carries another kind of pulse. In the nonphysical world, the spread of information is somewhat analogous to a pulse. The disturbance due to a pulse does not last long at any one spot, but the pulse travels on, carrying energy with it.

Sustained waves are emitted by vibrating bodies. When such waves pass through space or some material medium, the disturbance cannot continue in one direction only but must alternate. The reversal in the direction of the disturbance is a built-in property of electromagnetic waves, while in the case of elastic waves (including sound) reversal occurs because of the tendency of a deformed elastic body to spring back to its original state. Gravity supplies the restoring force for deep-sea water waves.

d | Nonperiodic versus periodic waves

In a *nonperiodic wave* the disturbance at a given point does not repeat itself in a regular manner and the waveform diagram does not show any repetitive pattern. Examples of such waves are noise and radio static.

In a *periodic wave* the disturbance is repeated at a regular interval called the *period T*. The number of vibrations passing a given point in unit time is called the *frequency f*. The distance at any instant between a point on the wave and the next successive point at which the whole wave pattern starts to be repeated is termed the *wavelength* λ (see Figs. 6–2 and 6–3). During the time T a wave crest moves forward to the position occupied by the preceding crest at the start of this time interval. The distance from crest to crest is the wavelength λ. Therefore the speed v with which a periodic wave is propagated forward is

$$v = s/t = \lambda/T. \hspace{3cm} \text{6–1}$$

Since $1/T = f$, we have

$$v = \lambda f. \hspace{3.5cm} \text{6–2}$$

This relationship, which follows solely from definitions, holds for all kinds of periodic waves. Since for many types of waves the speed v is well known, Eq. (6–2) makes it possible to compute λ when f is known, or f when λ is known.

Example 1. The speed of sound in air at 0°C is 331 m/sec, or 1087 ft/sec. Find the wavelength of a periodic sound wave in air at 0°C if the frequency of the wave is 500 cycles* per second.

Solution. From Eq. (6–2) we have

$$\lambda = \frac{v}{f} = \frac{331 \text{ m/sec}}{500 \text{ cycles/sec}} = 0.662 \text{ m/cycle} = 2.17 \text{ ft.}$$

Example 2. A radio station emits radio waves whose wavelength is 250 m/cycle. What is the frequency?

Solution. As stated above, $v = 3 \times 10^8$ m/sec for all radio waves. Therefore

$$f = \frac{v}{\lambda} = \frac{3 \times 10^8 \text{ m/sec}}{250 \text{ m/cycle}} = 1.2 \times 10^6 \text{ cycles/sec.}$$

This frequency may be expressed as 1.2 megacycles/sec (1 megacycle/sec = 10^6 cycles/sec), or as 1200 kilocycles/sec, which is a frequency in the AM (amplitude modulated) broadcast range.

For your information, the complete electromagnetic spectrum is outlined in Table 6–1. (Some of the ranges are being extended further.)

Two other important concepts related to *periodic* waves are amplitude and phase. By *amplitude* we mean the maximum magnitude of the disturbance produced by the wave. For the wave shown in Fig. 6–5 the disturbance is greatest at A_1, A_2, A_3, A_4, etc.; if the horizontal line represents the state of no disturbance, the amplitude of the wave is a, that is, the disturbance varies from a in the positive sense to a in the negative sense.

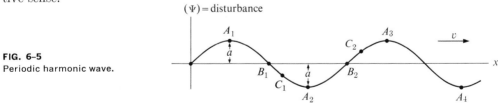

FIG. 6–5
Periodic harmonic wave.

---

* The word "cycles" refers to what is being *counted;* it is not part of the system of units. One may also speak of "vibrations per second," or "waves per second," but the words "cycles," "vibrations," etc., may be omitted if it is understood what is being counted.

TABLE 6-1

The Electromagnetic Spectrum

Type of radiation	Source	Approximate frequency range, cycles/sec	Approximate wavelength range
Micro pulses	Celestial	0.01–1	3×10^{10} m–3×10^8 m
Radio Long waves AM broadcast FM broadcast TV broadcast Radar	Electric circuits	10^4–5×10^5 5×10^5–1.6×10^6 88×10^6–108×10^6 55×10^6–900×10^6 10^9–10^{12}	3×10^4 m–600 m 600 m–188 m 3.4 m–2.8 m 5.5 m–0.33 m 30 cm–0.3 mm
Atomic Infrared Visible Ultraviolet X-rays	Excited molecules or atoms	10^{12}–3.8×10^{14} 3.8×10^{14}–7.5×10^{14} 7.5×10^{14}–10^{17} 10^{17}–10^{21}	0.3 mm–8000 Å* 8000 Å–4000 Å 4000 Å–30 Å 30 Å–0.003 Å
Nuclear Gamma rays	Radioactive nuclei	To above 3×10^{22}	To below 10^{-4} Å

*1Å = 1 angstrom = 10^{-10} m.

    The term *phase* is used with the same connotation as when we speak of the phases of the moon, or the phase our economy has reached in a business cycle. Points on a wave that are a wavelength apart (measured in the direction the wave is moving) are, from the definition of wavelength, points of equal disturbance, or points whose respective disturbances are "in phase." Points 2, 3, 4, . . . wavelengths apart are also points of equal phase. The symmetric waves that have the shape represented by $\Psi = a \sin x$,* as shown in Fig. 6–5, are ones of common occurrence; they are called *harmonic waves*. For such waves two points on the wave whose x-coordinates differ by half a wavelength, such as A_1, and A_2, B_1 and B_2, C_1 and C_2, are pairs of points for which the disturbances are half a cycle out of phase or, as we say, "completely out of phase." However, B_1 and C_1 represent a pair of points whose disturbances are only a little out of phase, A_1 and B_1 points whose disturbances are a quarter of a cycle out of phase, and so on.

---

    * $\sin x$ means "the sine of x." See Appendix 4.

As time passes, these relative phases do not change. Thus when the next crest reaches B_1 the next trough will reach B_2, and a crest and a trough are completely out of phase.

In the case of waves in elastic media the force F required to produce the disturbance Ψ is proportional to Ψ, so that $F = k\Psi$. The average force needed to increase Ψ from 0 to a is $\frac{1}{2}ka$, so that for the displacement $s = a$ the work done and energy E transferred are given by

$$E = Fs = \tfrac{1}{2}ka^2. \qquad\qquad \textbf{6-3}$$

Thus the *square of the amplitude* of such waves *is a measure of the energy* being transferred through the medium by the wave. This relationship holds for most periodic waves.

6|3 OBJECTIVE VERSUS SUBJECTIVE APPROACHES TO SOUND AND LIGHT

When trying to understand the difference between the scientific and the nonscientific point of view, we find it instructive to compare the way in which a physicist describes a sound or a colored light with the description an artist would give of the same sound or light. The scientist is first of all objective, while the artist is subjective. Furthermore, the artist is concerned with human emotions, and these in turn are related to the physiological and psychological makeup of man. What we hear depends both on the sound striking our ears and on the structure of our ears; light will produce a different sensation when seen by a colorblind person than when seen by one with normal eyes.

In the case of a sound produced, say, by a musical instrument, the description given by the physicist would consist of stating the intensity and the harmonic analysis of the wave. By *intensity* is meant the energy per second reaching a unit area placed at right angles to the direction of propagation of the wave. We saw that for all elastic waves the intensity is proportional to the square of the amplitude. *Harmonic analysis* means a description of the component frequencies present and of the amplitude of each. Thus in Fig. 6–6 the wave W is the result of the superposition of two harmonic waves, one with twice the wavelength, half the frequency, and nearly three times the amplitude of the other.

For a musician, the important ingredients of a musical sound are its loudness, its pitch, and its tonal quality, or timbre. *Loudness* is not the same as intensity because loudness refers to the relative response of the ear. If the intensity is increased from a value I to $10I$, then to $100I$, and finally to $1000I$, the ear interprets these three unequal increases in intensity as equal increases in loudness; in other words, the response of the human ear is logarithmic, and the decibel scale of loudness is based on this fact: each of the intensity increases mentioned above corresponds to an increase in loudness of 10 decibels (db).

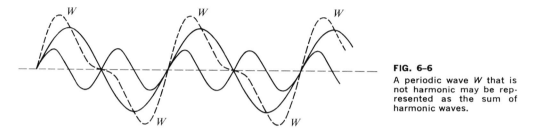

FIG. 6–6
A periodic wave *W* that is not harmonic may be represented as the sum of harmonic waves.

Pitch is the frequency the ear identifies with a musical sound. It is usually, though not always, the frequency of the lowest harmonic component.

Tonal quality depends on what harmonic components are present and the intensity of each. However, doubling the intensity of all the components may change the tonal quality, because the ear's response may reach a saturation point when the intensity of one frequency is doubled but not when the intensity of another frequency is doubled. It is, of course, in tonal quality that the sound of the note A from a violin differs from the sound of the note A from a piano. The value of a fine violin rests on the timbre of the sounds it emits.

Light is described by the scientist in terms of intensity, wavelength, and (if more than one wavelength is present) spectral distribution. We shall see presently how light waves of different wavelengths are separated out in different directions, or *dispersed*, by a prism or a ruled grating and the wavelengths present measured. In this way one may compare the output of an incandescent light bulb with that of a fluorescent lamp and show that the former emits a larger percentage of its energy in the wavelength range around 7000 Å (red light) than does the latter.

A nonscientist does not ask what wavelengths are present in the light from some source, but he is interested in its brightness and color. That *brightness* is not the same as intensity is obvious when one realizes that ultraviolet light, even when intense, is "dark light" to the eye. The normal eye is sensitive to a very narrow band of the electromagnetic spectrum, a band extending from about 4000 Å to 8000 Å, for which the ratio of the respective frequencies is only 2 to 1; this corresponds to but one octave of the musical scale.

By *color* we mean that quality of visible phenomena which we associate with red, green, blue, etc. Light of a single frequency will have a spectrum color, that is, a color found in the rainbow, or the spectrum of white light. Average daylight, often referred to as "white light," contains all of the visible spectrum colors in definite proportions. All possible colors may be made by mixing two or more colors in various proportions. The eye, however, does not analyze color this way, but it has the remarkable ability of combining colors to form new sensations; nevertheless, although one person can match two colors, it is doubtful if two people looking at the same color receive the same sensation. The ear, in contrast, is more analytical and can recognize two or more different frequencies at the same time, thus making possible the enjoyment of harmony and the complexities of contrapuntal music.

6|4 WAVE PROPAGATION IN THREE DIMENSIONS

Let us consider some of the properties of waves as they move through space, encounter obstacles, and enter different media.

Suppose that a wave travels out from a source and reaches all points on a certain surface at the same time t. Then this surface is one of constant phase for a given wavelength, and it is called a *wavefront*. Lines normal to wavefronts, or *rays*, represent the direction in which a wave is proceeding at any point.

Three-dimensional waves, regardless of type, all show the following properties: spreading out, reflection, refraction, diffraction, and interference. These will now be described.

In a homogeneous medium a curved wavefront spreads out as it advances, but retains the same shape.

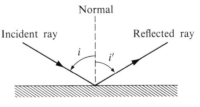

FIG. 6–7
Law of reflection. $\angle i = \angle i'$.

When a wave encounters a new medium, *reflection* occurs to a greater or less extent depending on various circumstances. It is found experimentally that the angle i between the incident ray and the normal to the interface equals the angle i' between the reflected ray and the normal (see Fig. 6–7); also, the two rays and the normal lie in the same plane. These statements constitute the *law of reflection*.

139

When a wave encounters another transparent medium, then, in addition to the reflected wave, a wave enters the second medium, but in the new medium the waves proceed in a different direction (Fig. 6–8). This bending of the rays and waves upon entering a new medium is called *refraction*. Let r be the angle between the refracted ray and the normal to the interface. It is found experimentally that at a given interface an increase in the angle i results in an increase in the angle r; however, the ratio i/r is not constant, as one might at first suppose. Further investigation shows that the ratio of the *sines* of the two angles is constant, or $\sin i \propto \sin r$.* Let us define the *index of refraction n* of the second medium relative to the first as

$$n = \sin i / \sin r. \qquad\qquad \textbf{6–4}$$

The statement that n is a constant for various angles of incidence constitutes *Snell's law of refraction*. This is one of the many restricted laws in physics; Snell's law is limited in its applicability and n is constant only for two given media. Values of n for various media relative to air may be found listed in tables.

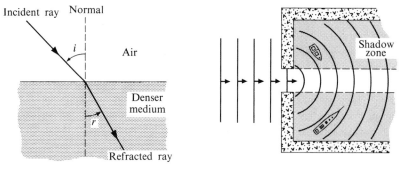

FIG. 6–8. Refraction.

FIG. 6–9. Diffraction of water waves passing into a harbor.

When a wave encounters an obstacle, so that part of the wave surface is cut off, effects result that are referred to as *diffraction*. One such effect is the apparent bending of the waves into the shadow zone behind the obstacle. A familiar example is furnished by sound; one can hear a person talking in the next room, or behind a screen, because the sound waves bend as they pass through a doorway or past the edge of a screen. Similarly, ocean waves bend around a breakwater into a harbor and cause boats moored behind the breakwater to bob up and down (Fig. 6–9).

---

* See Appendix 4 for the definition of the sine.

Finally, when waves from a source S (or from two synchronous sources) reach a point P via two different paths, SAP and SBP (Fig. 6–10), then the two sets of waves reinforce each other and produce a greater disturbance at P (than would either wave alone) if the two path lengths differ by a whole number of wavelengths, or

$$|\overline{SBP} - \overline{SAP}| = m\lambda, \qquad m = 0, 1, 2, 3, \ldots \qquad \textbf{6–5}$$

This is called *constructive interference*. On the other hand, *destructive interference*, or a lessening of intensity at P, results when the two paths differ in length by $\lambda/2$, $3\lambda/2$, $5\lambda/2$, etc.

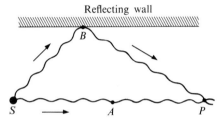

FIG. 6–10. Interference.

6|5 HUYGENS' PRINCIPLE OF WAVE PROPAGATION. FUNDAMENTAL LAW VI

The properties of waves discussed in the last section are independent bits of experimental knowledge. Nevertheless, scientists are always on the lookout for a unifying postulate, or principle, that will tie together many facts. (It is the aim of this book to set forth these fundamental postulates of physics in a clear and prominent fashion.) It should not seem unreasonable to look for a single principle that will explain all general wave propagation phenomena. Such a principle was first proposed by the Dutch physicist Huygens (1629–1695). His postulate may be stated as follows.

Every point on a wavefront may be regarded as a source of a secondary disturbance, which spreads out in the forward direction with the speed v of the wave at that point; after a time t these secondary wavelets will have radii vt and the envelope of these wavelets will be the new wavefront at the end of the time interval t.

This is essentially our *Fundamental Law VI*.

Of course this is not the same sort of law as Newton's law of gravitation or the conservation of energy principle. Huygens' principle does not postulate the existence of a fundamental force and its relationship to other physical quantities, nor does it state a general principle relating to

141

the constancy of some quantity. Our new law is simply a model for the correct construction of reflected, refracted, diffracted, and spreading waves. Physicists do not believe that secondary wavelets *really* radiate from every point on a wavefront, but we shall see that drawing such wavelets furnishes a useful working method for describing and predicting the facts associated with the propagation of waves. These facts must be covered by any set of fundamental postulates that tries to describe completely the behavior of our physical world. Since Huygens' principle does summarize the above-mentioned facts, and since it is perfectly general (it applies to any type of wave in any medium), it is included in our set of fundamental principles.

A wavefront is by definition a surface of constant phase and so the secondary wavelets we imagine spreading out from it will, at a given moment, all have the same phase. The envelope of these wavelets is the surface whose every point lies on one such wavelet, and hence it will also be a surface of constant phase, or a wavefront. As shown in Fig. 6–11, Huygens' principle immediately accounts for the spreading of a wave and the retention of its earlier shape in a homogeneous medium.

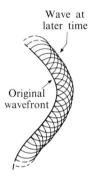

Wave at
later time

Original
wavefront

FIG. 6–11
A spreading wave retains its general shape, according to Huygens' principle.

Since Huygens' principle also made some erroneous predictions, it was subsequently modified and extended by Fresnel and others. Fresnel pointed out that when a set of waves of a single frequency falls on a slit, then every point on the slit may be regarded as a source sending out waves into the region beyond the slit; the disturbance at some point P behind the slit may be computed by adding the contributions of the wavelets reaching P at a given instant; but in making this addition, one must consider the relative phases of the contributing wavelets (see Section 6–8). We shall now show that Huygens' principle accounts for reflection, refraction, diffraction, and interference.

6|6 REFLECTION OF WAVES

The law of reflection may be derived from the modified Huygens' principle. Consider a plane wavefront ABC which is just reaching a reflecting surface MM' at the time t, as in Fig. 6–12. Suppose that at the time $t + \Delta t$ a wavelet from C reaches the point C', CC' being perpendicular to ABC. If the speed of the wave is v, then the distance $\overline{CC'}$ must equal $v \, \Delta t$. Now consider a wavelet spreading out from the midpoint B of the original wavefront. It will reach the reflecting surface at P at the time $t + \Delta t/2$; if a wavelet starts to spread out from P at this moment, it will attain a radius of $v \, \Delta t/2$ by the time the wavelet from C reaches C'. A similar analysis may be applied to wavelets originating at other points on ABC and producing secondary wavelets at various points on MM'. Finally, since the original wavefront has reached the mirror at A at the time t, the wavelet spreading out from A will have a radius of $v + \Delta t$ at the time $t + \Delta t$. The secondary wavelets referred to here are all in phase; hence their envelope $A'B'C'$ will be the new wavefront at the later time $t + \Delta t$.

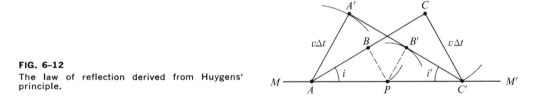

FIG. 6–12
The law of reflection derived from Huygens' principle.

In the right triangles ACC' and $AA'C'$, the side AC' is common and the sides CC' and AA' are each equal in length to $v \, \Delta t$. Therefore these triangles have equal third sides (from the Pythagorean theorem) and are equal triangles. Hence the reflected wavefront $A'C'$ makes the same angle with the mirror as does the incident wavefront AC. Since rays are perpendicular to wavefronts and the normal is perpendicular to a surface, the angle $CAC' = i$ and the angle $A'C'A = i'$. Hence $i = i'$, or the angle of incidence equals the angle of reflection.

Application of Huygens' principle to curved surfaces will show that when a plane wave is reflected from a *concave* mirror, as in Fig. 6–13(a), the reflected wave will *converge* toward a point F called the *focus* of the mirror. On the other hand, if a plane wave strikes a *convex* mirror, as in Fig. 6–13(b), the reflected wave will *diverge* away from the point F'. In general, reflection from a concave mirror makes converging waves converge more and diverging waves diverge less, or perhaps converge, while

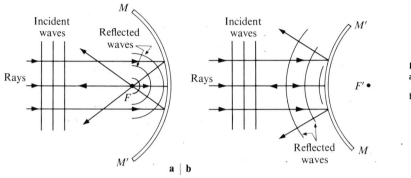

FIG. 6–13
a| Reflection from a concave mirror.
b| Reflection from a convex mirror.

convex mirrors produce the opposite effect. A continuation of this line of reasoning leads one to the theory of image formation by mirrors.

6|7 REFRACTION OF WAVES

Let us refer to Fig. 6–14 and derive Snell's law of refraction from the modified Huygens' principle.

The surface SS' separates medium 1, in which the speed of the waves is v_1, from medium 2, in which the speed is v_2. Here we make no assumption as to whether v_2 is or is not equal to v_1. The original wavefront at the time t is ABC. Let a wavelet from C reach the surface SS' at C' in the time Δt, so that $\overline{CC'} = v_1 \Delta t$. A wavelet from the midpoint B of AC will reach the surface at P in the time $\Delta t/2$, and so from P a secondary wavelet will spread out during the remainder $\Delta t/2$ of the time interval Δt. In medium 2 the wavelet from P will attain a radius of $v_2 \Delta t/2$ by the time the wavelet from C reaches C'. From A, a wavelet of radius $v_2 \Delta t$ will spread out into medium 2 in the time Δt. The new wavefront is thus $A'B'C'$.

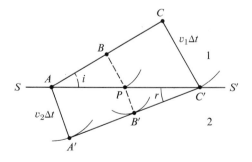

FIG. 6–14
The law of refraction derived from Huygens' principle.

In the right triangle ACC', the angle CAC' is i, the angle of incidence, and

$\sin i = v_1 \Delta t / \overline{AC'}.$

In the triangle $AA'C'$, the angle $AC'A'$ is r, the angle of refraction, and

$\sin r = v_2 \Delta t / \overline{AC'}.$

If we divide one equation by the other, we obtain

$$\frac{\sin i}{\sin r} = \frac{v_1}{v_2}. \qquad \textbf{6-6}$$

Since the speed of a wave in a noncrystalline medium is independent of its direction of motion, the ratio v_1/v_2 is independent of i and r for two given media. This is Snell's law.

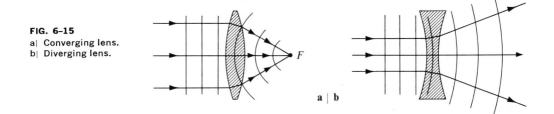

FIG. 6-15
a| Converging lens.
b| Diverging lens.

a | b

Note that Eq. (6-6) tells us something new, namely that when r is less than i, in which case $\sin r$ is less than $\sin i$, then v_2 must be less than v_1. Thus if rays are bent toward the normal upon entering a new medium, the speed of the waves should be less in the second medium. In general, if rays bend on entering a new medium, the speed v must change. The converse statement is also true. This relationship between speed and direction has been verified experimentally.

The fact that light travels more slowly in glass than in air accounts for the converging and diverging properties of various kinds of lenses. If a lens is *thicker in the middle,* rays passing through the middle will be slowed down more than those passing through the edges of the lens and the lens will increase the convergence or decrease the divergence of the waves striking it (Fig. 6-15a). A lens that is *thinner in the middle* will have the converse effect and so is called a diverging lens (Fig. 6-15b). Let us consider other illustrations of refraction.

Illustration 1

Figure 6–16 shows a ray of light striking a slab of *plate glass* with parallel sides. Upon entering the glass the light is bent toward the normal, or in a clockwise sense as viewed in the diagram. When passing back into the air, the ray is bent away from the normal, or counterclockwise in the figure. Since the angle r is the angle between the ray in the glass and the normal to *either* surface, the final direction of the ray is the same as its original direction, although the ray is slightly displaced sidewise.

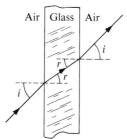

FIG. 6–16
Refraction by plate glass.

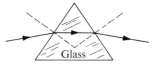

FIG. 6–17
Refraction by a prism.

Illustration 2

Figure 6–17 shows the bending of a ray of light by a *glass prism*. Again the light enters the glass and the ray is bent toward the normal; then the light leaves the glass and the ray is bent away from the normal. Why is the bending in the figure clockwise *each* time?

The deviation (change in direction) produced by a prism is found to be greater for blue than for red light, and the two colors are *dispersed* in different directions. (Which must go faster in glass?) Because the deviation is different for every wavelength, a prism may be used to separate out the colors present in white light, or light from any source, into the spectrum of that light.

Illustration 3

Figure 6–18 shows a pool of water of depth h. From the point P on the bottom of the pool, rays travel upward and strike the water-air surface at various angles of incidence i. For the ray labeled a, $i = 0°$, and the ray emerges unbent into the air. Ray b has a small angle of incidence i in the water, and this ray is bent away from the normal as it passes into the air because the speed of light in air is greater than in water. To a person in the air above the point P, the rays a and b will seem to diverge from the point P', the apparent position, or *image*, of P. The apparent depth h' will thus be less than the true depth of the water.

As the angle of incidence i in the water increases, the angle of refraction r in the air also increases, angle r always being greater than angle i, so that angle r will reach 90° before angle i does. The value of the angle i for which the angle r equals 90° is called the *critical angle*. In Fig. 6–18 ray c represents the one whose angle of incidence is the critical angle, so that it is bent 90° from the normal to the surface and passes along the interface between the water and the air. For rays such as ray d, whose angle of incidence in the water exceeds the critical angle, the water surface acts like a totally reflecting mirror and sends all of the light back into the water.

FIG. 6–18
Refraction of rays passing from water to air.

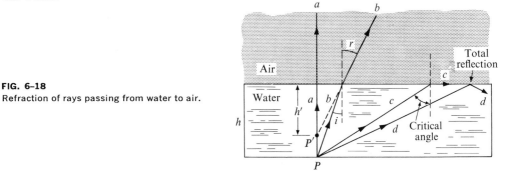

Since rays of light are reversible, that is, light may travel either way along them, the light from the sky passes into the eye of a fish in the form of a circular cone.

6|8 DIFFRACTION AND INTERFERENCE OF WAVES

According to Huygens' principle, when waves pass through an opening, every point in the opening may be regarded as the source of new wavelets. These wavelets will spread out in all directions in the region beyond the slit. This then explains why sound waves seem to bend as they pass through doorways and why water waves will bend around a breakwater. This bending, which is an example of *diffraction*, is not due to a change in the speed of the waves, but is an inherent property of waves in general.

Francesco Grimaldi (1618–1663), a professor of mathematics at the University of Bologna, Italy, made the first observations of the diffraction of light. In a narrow beam of sunlight he placed an opaque screen with a very small hole through which the light passed to a distant wall. Later he substituted for the screen a tiny obstacle. He observed that with the screen plus hole the spot of light on the wall was *larger* than expected,

147

while with the tiny obstacle the shadow on the wall was *smaller* than the obstacle. In both cases the effects are equivalent to a slight bending of the light around whatever obstacle it passes. Grimaldi also noticed narrow color fringes near the edge of the light pattern on the wall. This was also partly a diffraction effect, and its explanation is related to the one we shall presently give of Young's color fringes produced by a double slit.

Thomas Young (1773–1829) made the discovery that light exhibits the property of interference. This was around the year 1800. Why was this discovery made so late when Newton and others had observed some time earlier the interference of two sets of water waves? The answer now given to this question is as follows. The wavelength of light in the visible region of the spectrum is *much less* than that of sound and water waves. When waves bend around an obstacle they pass into the geometric shadow zone determined by the line drawn from the source of the waves to the edge of the diffracting object (see Fig. 6–9). Both experiment and a more detailed theory show that the intensity of the diffracted waves falls off rapidly as a function of distance into the shadow zone if this distance is expressed in terms of number of wavelengths; for a sharp obstacle the intensity is appreciable for distances into the shadow zone of only a few wavelengths. Thus sound can be heard coming from a source that is hidden from sight, say by a tree or the corner of a building (Fig. 6–19).

FIG. 6–19
While sound waves (with wavelengths of the order of a meter) diffract around corners, visible light (with wavelengths of around 6×10^{-7} m) travels out from the same source along rays that are not noticeably bent.

When observing diffraction and interference effects with slits, it is necessary to make the slit width the order of a wavelength of the waves to be diffracted. Young made two very narrow slits A and B (see Fig. 6–20) close together (call the separation d) on a screen. These slits were placed equidistant from a source S, such as a sodium lamp, that emitted a single wavelength λ of light. The light waves in such an experiment reach each slit in phase. The slits themselves may be regarded as synchronous sources of two new sets of waves of wavelength λ. It is important when one wishes to obtain interference effects with light that the two sets of overlapping waves come from a common source. This is because the wave trains from a light source are not of infinite length and abrupt changes in phase occur from one wave train to the next. For the interference of sound waves and the ripple waves in a pond or tank, one may use synchronous vibrators to produce the two sets of waves.

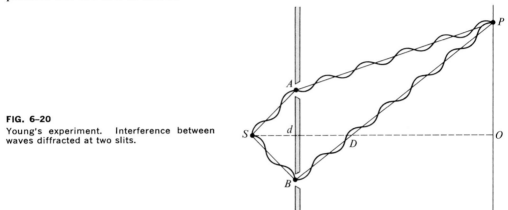

FIG. 6–20
Young's experiment. Interference between waves diffracted at two slits.

Returning to Young's experiment, the two sets of waves, spreading into the region behind the slits, overlap one another. At a point P on a second screen, placed at a distance D behind the first, the condition for constructive interference is given by Eq. (6–5), namely

$$|\overline{SBP} - \overline{SAP}| = m\lambda, \qquad m = 1, 2, 3, \ldots$$

This equation is based on the idea of *superposition*. It is found that when two sets of waves pass across the same point the disturbance is the sum of the disturbances produced by each wave separately. So if the crest of one wave coincides with that of another, the disturbance will be a maximum in the upward sense, soon to be followed by a maximum downward disturbance as the succeeding troughs of the two wave trains pass across the point under discussion. Should, however, a crest meet a trough, the disturbance will be lessened and possibly (as for waves of equal magnitude) reduced to zero. In Fig. 6–20 the distance $\overline{SAP} = 6.5$ wavelengths, the distance $\overline{SBP} = 7.5$ wavelengths, and two crests are about to meet at P; this represents *constructive* interference, with $m = 1$. However, for $\overline{SAP} = 6.5$ wavelengths and $\overline{SBP} = 8.0$ wavelengths we would have maximum *destructive* interference at P.

The point O that is on the second screen and on the perpendicular bisector to AB is the point equidistant from the two slits, or the point for which $m = 0$. Hence, O will be a point of maximum brightness.   **149**

Let P be the point, above or below O, for which $m = 1$. This will be the next point on either side of O where maximum brightness will be obtained. Let y be the distance from O to P. As in Fig. 6–21, draw AC perpendicular to BP. Since $m = 1$ for the point P, the distance $\overline{BP}$ must be one wavelength λ longer than the distance $\overline{AP}$. The difference between these two distances is essentially the distance $\overline{BC}$ in the figure. (We assume that the second screen is many wavelengths behind the first.) The angle BAC is essentially equal to the angle PMO, called θ, because their respective sides are (or nearly are) perpendicular. Thus we may consider triangles ABC and PMO to be similar right triangles. Hence by proportion of corresponding sides we have

$$\frac{\overline{BC}}{\overline{AB}} = \frac{\overline{OP}}{\overline{MP}}.$$

Since y is much less than D (its size is reduced in the figure), we may take $\overline{MP} = D$. With $\overline{BC} = \lambda$, $\overline{AB} = d$, and $\overline{OP} = y$, our proportion becomes

$$\frac{\lambda}{d} = \frac{y}{D}. \qquad\qquad \textbf{6-7}$$

This formula may be used to compute λ when d, y, and D are known (see Experiment 10 at the end of this chapter).

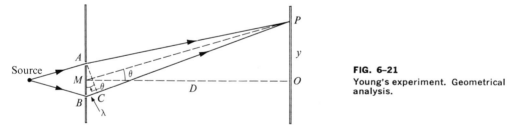

FIG. 6–21
Young's experiment. Geometrical analysis.

If the slits are long and narrow, one will obtain a bright fringe through O, a parallel one through P, and others, about equally spaced, on beyond. This is called an *interference pattern*.

Example. Suppose that $d = 0.2$ mm, $D = 1$ m, and $y = 2.5$ mm when $m = 1$. What is the value of λ?

Solution. Express d and y in meters. Then $d = 2 \times 10^{-4}$ m and $y = 2.5 \times 10^{-3}$ m, so that

$$\lambda = \frac{yd}{D} = \frac{2.5 \times 10^{-3} \text{ m} \times 2 \times 10^{-4} \text{ m}}{1 \text{ m}} = 5 \times 10^{-7} \text{ m}.$$

This is about the wavelength of green light. It is remarkable that a quantity as small as the wavelength of visible light may be measured by this rather simple experiment.

Since the intensity of diffracted light is low, the use of a large number of equally spaced slits is an improvement over Young's two-slit experiment. A ruled *grating* may contain some 15,000 lines per inch, which is equivalent to the same number of slits, each of which contributes to the diffracted light. The condition for constructive interference is the same as for the two slits, namely that the additional path length of each successive ray must be equal to a whole number of wavelengths, that is, $\overline{BC} = m\lambda$, $m = 0, 1, 2, 3, \ldots$ Equation (6-7) then becomes

$$\frac{m\lambda}{d} = \frac{y}{D}, \qquad\qquad \textbf{6-8}$$

where y has a different value for $m = 1$ (called the *first-order* spectrum), for $m = 2$ (the *second-order* spectrum), etc. For larger values of m (and hence of y) Eq. (6-8) is only approximately correct and it is better to use the more exact relation (see triangle ABC of Fig. 6-21, in which $\overline{AB} = d$ and $\overline{BC} = m\lambda$)

$$\frac{m\lambda}{d} = \sin\theta. \qquad\qquad \textbf{6-9}$$

where d is the separation of the lines ruled on the grating, θ is the angle between the diffracted rays and the perpendicular to the grating, and m is the order number. Note that when $m = 0$, $\theta = 0$ for *all* wavelengths, while when $m = 1$ the value of θ increases with the wavelength. Longer waves of red light are diffracted more than shorter waves of blue light; again we have a separation of different colors, or *dispersion*.

6|9 THEORIES OF LIGHT

The attempts of physicists to explain the properties of light illustrate well the nature of physical theory.

In the time of Newton (1642-1727) it was known that light seemed to travel in straight lines, and that light rays obeyed the laws of reflection and refraction. The diffraction and interference properties of light had not yet been discovered. In view of this state of the knowledge about light, Newton proposed a *corpuscular theory* of light which adequately explained the known facts. He postulated that a beam of light consists of a stream of fast-moving tiny particles, somewhat like the stream of bullets from a machine gun. This immediately explained the apparent sharp shadows cast by obstacles to a beam of light. Reflection could be explained as due to elastic collisions between the light particles and the reflecting surface. To explain refraction, Newton had to postulate that the light particles are attracted by glass and water, so that when a light particle approaches, say, a glass surface from air, the normal component of the velocity increases and the other components do not change. Thus Newton drew a new conclusion from his theory, namely that light must go *faster* in glass than in air. The inability of seventeenth and eighteenth century physicists to measure the speed of light in glass or water left a situation in which Newton's theory was completely adequate and even superior to the wave theory, in that Newton's theory extended the well-known laws of mechanics to light.

With the discovery of diffraction and interference, Newton's theory of light became inadequate. It did not "grow gracefully" and furnish an explanation of these new effects, whereas the wave theory did. The wave theory predicted that the speed of light would be found to be *less* in glass than in air, which is the opposite of Newton's prediction. Finally, in 1850, Foucault succeeded in showing experimentally that the wave theory's prediction was the correct one.

In the twentieth century further facts about light have been discovered. These facts concern the emission and absorption of light. Since the wave theory attempts to explain only the propagation properties of light, these new facts do not invalidate the wave theory; rather they call for an additional theory that explains how light is emitted and absorbed. This additional theory is the quantum theory, which we shall discuss later. In summary we may say that Newton's theory was adequate for his day, but it did not prove fruitful and new facts could not be explained by it.

PROBLEMS

1. Suppose that at $t = 0$, $\Psi = 0$, except between $x = 1$ m and $x = 1.5$ m where $\Psi = 5$ units, and that this pulse is moving in the x-direction with a speed $v = 0.5$ m/sec. Plot Ψ against x at $t = 0$, 1, and 2 sec, respectively.

2. A wave like that shown in Fig. 6–5 is represented by a wave function such as $\Psi = A \sin (kx - bt)$. This is called a sinusoidal or *harmonic wave*. Here A, k, and b are constants for the wave. Plot Ψ vs. kx at

a) $t = 0$,   b) $bt = \pi/2$,   c) $bt = \pi$.

d) Which way is the wave traveling? What is its amplitude, its wavelength, and its speed? (Express answers in terms of A, k, and b.)

3. Find the frequency of

a) a radar wave of 10-cm wavelength, and

b) a green light of 5000-Å wavelength.

4. The Greek scientist Pythagoras discovered that a note of frequency f_1 and a note of frequency f_2 sounded together will make a chord that is pleasing to the ear if the ratio $f_2/f_1 = m/n$, where m and n are different integers between 1 and 6. In the diatonic musical scale, the frequencies of the eight notes relative to do are: do 1, re $\frac{9}{8}$, mi $\frac{5}{4}$, fa $\frac{4}{3}$, sol $\frac{3}{2}$, la $\frac{5}{3}$, ti $\frac{15}{8}$, do 2. Find all the pleasant two-note chords obtainable from these eight notes.

5. Make up a six-note scale such that the highest note has twice the frequency of the lowest and for which there exist a maximum number of pleasant two-note and three-note chords.

6. Look up and describe the *chromaticity* diagram for colors. What are *complementary* colors?

7. Show that in reflection from a plane mirror the image is as far behind as the object is in front of the mirror.

8. a) How long must a mirror be for a 6-ft man to see his full length in it?

b) In what respect do you see yourself in a mirror differently from the way your friends see you face-to-face?

9. Show that when a mirror is turned through θ degrees, light reflected from the mirror is rotated through 2θ degrees.

10. Show that the sine of the critical angle is the reciprocal of the index of refraction of the denser medium relative to the lighter medium.

11. A glass prism whose angles are 45°, 45°, 90° has a critical angle of 42°.

a) How could this prism be used to reflect light and change its direction by 90°?

b) by 180°?

12. Suppose that Young's experiment is done with water waves (ripples) of 1-cm wavelength and $d = 6$ cm. Find the distance between points of reinforcement at a distance $D = 120$ cm.

13. Suppose that a grating has 5000 lines per centimeter.

a) From Eq. (6–9) find θ for $m = 1, 2,$ and 3 and $\lambda = 4.4 \times 10^{-7}$ m (blue light).

b) Find θ for $m = 1, 2,$ and 3 and $\lambda = 6.6 \times 10^{-7}$ m (red light).

c) Show on a diagram the relative directions of the three diffracted blue waves in (a) and the three diffracted red waves in (b).

14. The colors of thin films are due to interference effects. In explaining these effects one must include the following facts. (1) Light waves undergo a phase change of half a cycle when reflected from the near surface of the film back into air, but not when reflected from the far surface back into the film. (2) Due to the change in speed, a distance d traveled by the light in the film is equivalent to a distance nd in air. If white light falls normally on a film whose $n = 1.5$, what color will the film have by reflected light if its thickness is

a) 1.1×10^{-7} m,

b) 7.5×10^{-8} m,

c) zero?

(For relationships between wavelengths and colors see problem 13.)

15. Look up and explain the conditions for standing waves in a string clamped at both ends.

Experiment 10 THE GRATING
Diffraction and Interference of Light

Object: To measure the wavelength of sodium light and to observe the diffraction of white light using a diffraction grating.

Apparatus: A coarse diffraction grating with about 250 lines per centimeter, if possible, slit, optical bench, sodium burner, micrometer microscope.

Procedure

Step 1. Place the sodium burner behind a V-shaped slit in a scale and observe through a grating in front of the scale the many orders of the spectrum (Fig. 6–22). Keeping your eye close to the grating, have your partner slide a piece of paper over the scale until its edge coincides with the middle of the first image. Record its position and repeat for at least the first five orders on either side of the central image. Record the distance D from the grating to the slit.

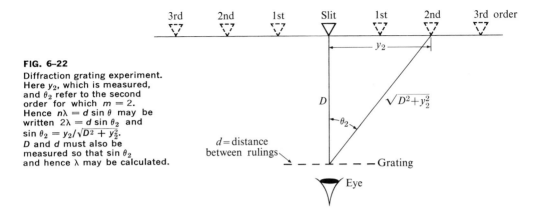

3rd   2nd   1st   Slit   1st   2nd   3rd order

FIG. 6–22
Diffraction grating experiment.
Here y_2, which is measured,
and θ_2 refer to the second
order for which $m = 2$.
Hence $n\lambda = d \sin \theta$ may be
written $2\lambda = d \sin \theta_2$ and
$\sin \theta_2 = y_2/\sqrt{D^2 + y_2^2}$.
D and d must also be
measured so that $\sin \theta_2$
and hence λ may be calculated.

d = distance
between rulings

y_2

D

$\sqrt{D^2 + y_2^2}$

θ_2

Grating

Eye

Step 2. Put an incandescent lamp in place of the sodium burner and record carefully the various characteristics of the resulting spectra. Draw labeled diagrams.

Step 3. Remove the grating and measure the grating space with a micrometer microscope. It is better to measure the distance between ten or more lines and find the average value of d. Why? Check your value by taking another reading.

Step 4. For each order in Step 1 average right and left readings and calculate the wavelength, using Eq. (6–9). Find your average value of λ for all the orders and also compute the mean deviation. Compare with the accepted value of 5893 Å for the wavelength of sodium light.

Step 5. Discuss the white light spectrum. Why does the red in the first orders appear to be a pure red, while in the second orders it is replaced by a pink or magenta color?

7

7|1 ELECTRIC CHARGE   Experiments on static electricity were performed by the Greeks. For example, they found that when amber was rubbed, it possessed the ability to attract small objects, such as a piece of thread. They offered no scientific theory to explain this phenomenon, but just stated that a "sympathy" existed between the amber and the attracted objects. Electrical effects related to the atmosphere (lightning, St. Elmo's fire, etc.) were, of course, known to ancient man, but no explanation of them was offered until modern times. The application of the experimental method to electrification phenomena and the development of a quantitative theory culminated in the work of the French scientist Coulomb (1736–1806).

The Conservation of Charge and Coulomb's Law of Electrostatic Force

Let us take a look at some basic experiments that led up to Coulomb's work. These may be repeated easily in the laboratory and the reader should try them, or at least witness their demonstration.

First let us take two small pith balls and suspend them near each other, using silk threads (which are nonconducting). Next take a piece of cat's fur and with it rub a rod of rubber or bakelite. Stroke each pith ball with the rod. The balls will fly apart, indicating that there is a force of repulsion between them (Fig. 7–1). It is evident that something was added to the balls to cause this repulsion and this "something" is called *electric charge*. Since each ball must have received the same sort of charge from the rod, we see that *like charges repel*.

Now discharge the balls by holding them in your moist hand (so that their charges will pass down through your body). If the balls are stroked with a glass rod rubbed with silk they will again fly apart, but if one is touched by the rubber rod and the other by the glass rod then the balls are attracted toward one another (Fig. 7–2). This indicates that the balls

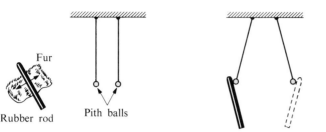

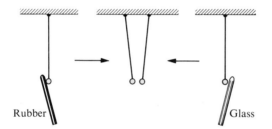

FIG. 7-1. Like charges repel.

now have different kinds of charge and that *unlike charges attract*. One may also note a slight attraction between the ball charged by the rubber rod and the cat's fur used to rub the rod, as well as between the ball stroked by the glass rod and the silk used to rub that rod.

FIG. 7-2. Unlike charges attract.

7|2 THE CONSERVATION OF CHARGE PRINCIPLE. FUNDAMENTAL LAW VII

The experiments just described suggest that there are at least *two* kinds of electric charge. There seems to be no need to postulate more than two kinds. An uncharged or *neutral* body may be thought of as one which possesses either no charge or equal amounts of the two kinds. This leads us to postulate that the two kinds are the opposites of one other, like plus and minus quantities which tend to cancel each other out. On the other hand one might assume that neutral bodies contain a normal amount of some sort of electric fluid and that positive electrification is simply due to an excess of electric fluid and negative electrification results from a deficit of the electric fluid. For a time there was debate as to whether there were two kinds of fluid or only one. The present view, based on the electron theory of matter, has proved to be the most fruitful and we shall adopt it. A brief description of this theory follows.

As mentioned earlier, matter is considered to be composed of atoms and each atom to consist of a core, or nucleus, surrounded by a cloud of orbiting electrons. Long before the discovery of the electron Benjamin

Franklin arbitrarily chose to take the charge left on glass rubbed with silk as positive and this convention has been followed ever since. When the electron was isolated just before 1900 it was found to have the other kind of charge, that is, the kind left on rubber stroked with fur, so we say that the electron is a negative particle. Atomic nuclei always have positive charges. Normally an atom is *neutral*, that is, the sum of the negative charges on the electrons is just equal and opposite to the positive charge on the nucleus. When two objects are rubbed together and one has a greater affinity for electrons than the other, then some electrons will be surrendered by the other. This will result in an excess of electrons, or a *negative* charge, on the object with the greater electron affinity and a deficit of electrons, or *positive* charge, on the other. It may seem that we have adopted a one-fluid theory, the fluid being negative electrons. However, there are important differences. We shall find that the charge $(-e)$ on the electron is finite and fixed, so electrons cannot constitute a continuous fluid. Furthermore, elementary particles (protons and positrons) have been discovered with positive charges of $+e$, hence positive and negative charge are equally fundamental, bringing us back to a two-kinds-of-charge theory.

We see that according to the electron theory a charge of one sign may be separated from a charge of the opposite sign, as by rubbing. Conversely, a negative charge may recombine with and neutralize a positive charge of equal magnitude. Recent work in high energy physics (Chapter 13) indicates that an electron and positron may actually be created or annihilated in pairs, but, since their charges just cancel, such processes do not change the net charge in the world either. Thus we are led to another fundamental conservation principle, the *law of the conservation of charge*, which postulates the following:

In natural processes positive and negative charges may be separated or recombined, but the net charge in our world remains constant.

This is our *Fundamental Law VII.*

Let us test this principle with another familiar experiment. Just as a comb that has been run through hair will attract other light objects, so will our charged rods tend to pick up small uncharged pieces of paper. How can a charged body attract an uncharged one? The explanation, illustrated by Fig. 7–3, is that the neutral body contains charges of both signs and that these separate to some extent. A negatively charged comb pulls the positive charges in the paper toward the side nearest the comb and repels the negative charges to the far side. If we make what seems a likely assumption, that the comb attracts the plus charges more than it repels the negative ones because the plus charges are nearer to the comb, then we see that the net force between the comb and paper must be attractive. Positive charge was not created in the paper, it was just separated from an equal negative charge.

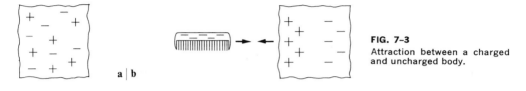

FIG. 7-3
Attraction between a charged and uncharged body.

a | b

Bodies in which charges are free to move are called *conductors*. There are all degrees of conductivity, running from the high conductivity of metals and electrolytes down through that of the so-called *semiconductors*, to that of the poorest conductors, which we label *insulators* (e.g., amber, silk, porcelain, and certain plastics). In metals and most other solids the charges which are free to move are electrons. A *free electron* is one that has become detached from, and therefore is no longer associated with, an atom. The atoms from which electrons have broken away become positively charged ions, which are heavy and much less free to move.

Just what is the nature of this thing called charge which, for example, electrons possess? All we can say is that charged bodies possess properties which uncharged bodies do not and then we can describe these properties. Every material body possesses the property of mass and as a result (1) it is subject to gravitational forces and (2) it shows reluctance to change its state of motion. Charged bodies are in addition (3) subject to electric and magnetic forces, the nature of which will be discussed in this and the next chapter.

7|3 COULOMB'S LAW OF ELECTROSTATIC FORCE. FUNDAMENTAL LAW VIII

Coulomb knew of the experiments just described. He recognized that just as gravitational forces are fundamental and universal, so the forces between charged bodies indicate the existence of another fundamental type of force in our world. Since this force exists between charged bodies at rest, it is called the *electrostatic* or *electric force*.

Being a first-rate experimental physicist, Coulomb next asked himself this question: On what factors does the electric force depend and how does it depend on each? He sought for the law relating the electric force to other physical quantities.

By means of a delicate torsion balance capable of measuring a force of less than 10^{-7} N (Fig. 7–4), Coulomb found that the force between two charged balls varied inversely as the square of the distance r between their centers. This relationship had been proposed by others, but Coulomb's work gave it a firm experimental foundation.

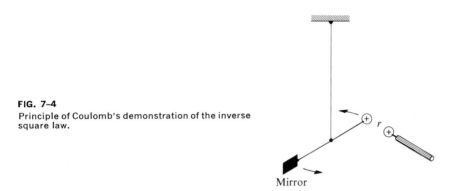

FIG. 7–4
Principle of Coulomb's demonstration of the inverse square law.

Mirror

Coulomb, realizing that the electric force between two charged bodies must also depend on their respective charges, q_1 and q_2, devised a way of determining the *relative* charge on bodies. He suspended a ball with a charge which we shall call q; he then touched this ball to an uncharged ball of the same size as the first so that the charge q became equally shared. Each ball then had a charge of $q/2$. By letting one of these balls share its charge with a third similar ball, charges of magnitude $q/4$ could be obtained, and so on. This gave him a set of charges of various magnitudes and made it possible for him to measure how the electric force depends on the charges on the interacting bodies. His results may be summarized as follows.

Bodies with like charges repel and those with unlike charges attract one another; for point charges (or small charged spheres) the force of interaction is proportional to the product of the charges and inversely proportional to the square of their distance apart.

Expressed mathematically, Coulomb's law states that the electric force F_e on a point charge q' due to the presence of another point charge q is

VIII. $$F_e = k_e \frac{qq'}{r^2},$$ **7-1**

where k_e is the constant of proportionality, r the distance of separation, a positive force means repulsion, and a negative force signifies attraction. This is our *Fundamental Law VIII.*

161

You will note the close resemblance between this law and Newton's law of gravitation. Coulomb and his immediate predecessors were undoubtedly guided by this analogy. The inverse-square-of-the-distance factor in both laws may be attributed to the geometric properties of Euclidean space in which anything that spreads out with spherical symmetry across successive spherical surfaces of radius r and area $4\pi r^2$ must become diluted in inverse proportion to such an area (Fig. 7–5). But it is not obvious that gravitational and electrical forces must be central and spherically symmetrical; it is just a fact that they are.

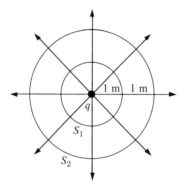

FIG. 7–5
Same total number of lines cross both surfaces S_1 and S_2, but since $S_2 = 4S$, the concentration, or lines per unit area, is one quarter as much for S_2 as it is for S_1.

In Newton's law of gravitation,

$$F_g = G\frac{m_1 m_2}{r^2},$$

we took the units of force, mass, and distance to have been defined, so that this equation defined the new quantity G; the value of G which makes the equation balance had to be determined experimentally, because it depends not only on our choice of units but also on the strength of *gravitational* forces. The strengths of the various forces in our world are inherent properties of our world. We shall see that the electric force is relatively a strong one. As with Newton's law, we shall take the units of force, charge, and distance to have been chosen and let Eq. (7–1) define k_e, the value of which must be found experimentally (see Section 7–8). Our unit of charge, the coulomb (abbreviated to C), will be defined in the next chapter, and we shall see that its definition is based on another fundamental law (Ampère's) and another type of force (magnetic). With F_e measured in newtons, q and q' in coulombs, and r in meters, the experimental value found in air or vacuum for k_e is almost, but not exactly,

162 $k_e = 9 \times 10^9$ N-m^2/C^2.   **7-2**

Example 1. Compare the electric force between an electron and a proton with the gravitational force between these particles for the same separation.

Solution. From experimental information we have $q = 1.6 \times 10^{-19}$ C and $m_1 = 9.1 \times 10^{-31}$ kg for the electron, and $q' = 1.6 \times 10^{-19}$ C, $m_2 = 1.6 \times 10^{-27}$ kg for the proton; $G = 6.67 \times 10^{-11}$ N-m^2/kg^2. Both forces are attractive and, since the r^2 terms cancel, the ratio of their magnitudes is

$$\frac{F_e}{F_g} = \frac{k_e}{G} \frac{qq'}{m_1 m_2}$$

$$= \frac{9 \times 10^9 \times 1.6 \times 10^{-19} \times 1.6 \times 10^{-19}}{6.67 \times 10^{-11} \times 9.1 \times 10^{-31} \times 1.6 \times 10^{-27}}$$

$$= 2.4 \times 10^{39},$$

which is one of the largest numbers in our physical world. The value of the ratio F_e/F_g varies with the masses of the elementary particles chosen; the ratio is about 10^{36} for two protons (heavy particles), 10^{38} for two mesons (particles of intermediate mass), and 4×10^{42} for two electrons. On the average it is about 10^{38}.

Example 2. Suppose that for a certain pair of charges, each of magnitude q, with a separation of 1 cm ($=10^{-2}$ m), the electric force of repulsion is 9×10^{-3} N. Find q in coulombs.

Solution. From Eq. (7–1) we have

$$9 \times 10^{-3} \text{ N} = 9 \times 10^9 \frac{\text{N-m}^2}{\text{C}^2} \times \frac{q^2}{10^{-4} \text{ m}^2},$$

$$q^2 = 10^{-16} \text{ C}^2,$$

$$q = 10^{-8} \text{ C}.$$

7|4 THE ELECTRIC FIELD

Since an electric force exists between charged particles located in even the best vacuum, such a force, like that of gravitation, furnishes an example of what has been called "action at a distance," and it cannot be given a mechanistic explanation. This point was one that the followers of Newton found hard to accept, and the reader may find it bothersome if he has not become accustomed to it. There is, however, another way to look at the situation, a way that has proved fruitful; it is to introduce the concept of a *field*. This was first suggested by the English physicist Michael Faraday (1791)–1867). The concept of fields plays an important role in modern physics, in which it is combined with the quantum theory (see Section 14–5).

Faraday proposed the following alternative statement of Coulomb's law, a statement that summarizes the same experimental facts.

a) *Around every charged body there exists an electric field* **E**, *which is a vector quantity.*

b) *The direction of the electric field is away from a positive charge and toward a negative charge.*

c) *The strength of the electric field at a distance r from a charge q, due to q, is*

$$E = \frac{k_e q}{r^2}.$$

7–3

d) *The force of an electric field* **E** *on a charge q′ placed in the field (but not contributing to it) is*

$$\mathbf{F}_e = q'\mathbf{E},$$

7–4

where **E** *represents the field at that point. The force is in the direction of the field if q′ is positive, opposite if q′ is negative.*

We see that (b), (c), and (d) together are equivalent to Coulomb's law. These statements are illustrated by Fig. 7–6.

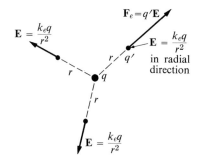

FIG. 7–6
Field around a point charge q exerts a force on a test charge q' placed in this field.

Faraday made the further postulate:

e) *If q′ is in the neighborhood of more than one charge, then* **E** *in Eq. (7–4) must be regarded as the vector sum of the electric fields due to each of the charges (other than q′).*

In other words, the electric field of one charge exists independently of that of any other charge. It is thus possible to compute the field due to any given distribution.

The concept of the field and especially the lines drawn to represent it (see next section) are, of course, another example of a useful model that helps us to picture and remember a lot of facts and to work out what will happen in new situations. It must again be emphasized that the lines do not actually exist and that what does have reality is the fact that neighboring charges interact. Later we shall see that the transmission of energy is associated with regions containing fields and energy is something that we can detect and measure.

One should note the reciprocal relationship between electric charge and field. A charge "produces" (has associated with it) an electric field and an electric field acts on any charge placed in the field. Equation (7–4) may be taken as the definition of $\mathbf{E}$ ($\mathbf{E}$ = force per unit charge), so that in the mks-coulomb system of units $\mathbf{E}$ must be expressed in newtons per coulomb (N/C). This ensures that in Eq. (7–4) $\mathbf{F}$ and $q'\mathbf{E}$ will both be stated in newtons and satisfies the rule that both sides of an equation must be expressed in the same units.

Example. Two charges of equal magnitude are situated on the x-axis, as shown in Fig. 7–7.

a) If the charges are both positive, find the direction of $\mathbf{E}$ at the points P, O, and R respectively.

b) Repeat for the case where q_1 is negative and q_2 positive.

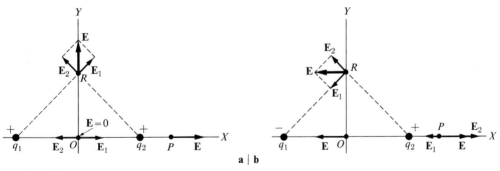

FIG. 7–7
a| Field due to two equal plus charges.   b| Field due to equal and opposite charges.

Solution. Let $\mathbf{E}_1$ be the field due to q_1 and $\mathbf{E}_2$ that due to q_2. Remember that the field is away from a positive charge and toward a negative one, and that fields add vectorially.

a) At P (Fig. 7–7a), $\mathbf{E}_1$ and $\mathbf{E}_2$ are both to the right (they are not shown in the figure) and therefore their vector sum $\mathbf{E}$ is to the right.   **165**

At O, $\mathbf{E}_1$ is to the right, $\mathbf{E}_2$ to the left. Also $\mathbf{E}_1$ and $\mathbf{E}_2$ are equal in magnitude, since the charges and distances are equal. Therefore $\mathbf{E}_1$ and $\mathbf{E}_2$ cancel each other and $\mathbf{E} = 0$.

At R, $\mathbf{E}_1$ and $\mathbf{E}_2$ have equal magnitudes and equal upward inclinations; hence they add to give an $\mathbf{E}$ along the y-axis, as in Fig. 7–7(a).

b) At P (Fig. 7–7b), $\mathbf{E}_1$ is now to the left, $\mathbf{E}_2$ to the right, so that one subtracts from the other. However, $\mathbf{E}_1$ is weaker than $\mathbf{E}_2$ because q_1 is farther away then q_2. Hence $\mathbf{E}_2$ predominates and $\mathbf{E}$ is to the right.

At O, $\mathbf{E}_1$ and $\mathbf{E}_2$ (not shown in figure) are both to the left and therefore $\mathbf{E}$ is to the left (and not zero).

At R, $\mathbf{E}_1$ and $\mathbf{E}_2$ have equal magnitudes and equal inclinations to the left, hence they add to give an $\mathbf{E}$ to the left, as in Fig. 7–7(b).

The above example is typical of the many in physics where simple substitution in a formula will not grind out the answer. Rather, one must study the "rules of the game," which, in this case, are (a) to (e) above, and play accordingly.

7|5 LINES OF FORCE

Lines of force are drawn to help us visualize the field at all points; they show us quickly the whole pattern of the field. A line of force is, by definition, a line or curve such that at any point on it the tangent to the curve is parallel to $\mathbf{E}$ at that point. Thus lines of force show us the *direction of the field* at many points. At points between lines of force we can usually guess the direction of the field by interpolation.

What makes the lines-of-force model of the electric field especially useful is the fact, mentioned earlier, that the field at a distance r from a point charge is proportional to $1/r^2$, whereas the area of a sphere with radius r, centered at the charge, is proportional to r^2. If we draw a fixed number of lines spreading out symmetrically from the charge, these lines become spaced farther and farther apart in just such a way that the concentration of the lines, or the number of lines per unit surface perpendicular to $\mathbf{E}$, is proportional to the *strength* of the field at any point. For example, see again Fig. 7–5. Thus by drawing a fixed number of lines per unit charge, leaving positive charges and ending eventually on equal negative charges, one obtains a picture of both the *direction and strength* of the electric field for a given charge distribution.

Illustration 1. *Electric field between parallel plates*

The use of parallel conducting plates makes it possible to produce a uniform electric field whose magnitude may be computed from measurable quantities.

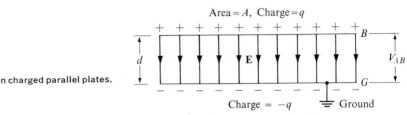

FIG. 7–8
Electric field between charged parallel plates.

Figure 7–8 shows a metal plate B separated by a distance d from a similar parallel plate G. Suppose that B is insulated and given a charge $+q$, while G is connected to the earth ("grounded"). The earth is a tremendous conducting sphere, and the positive charge on B will attract electrons up from the earth onto G, which is as near as these electrons can get to B. If the plates are close together compared with their diameters, the charges on the two plates will be on the surfaces facing one another. There can be no field inside conductor B, for, if there were, it would cause electrons in B to move until all fields were canceled, and we would then have the static situation that has been assumed. Therefore every line of force leaving B will end on G. Lines leaving a charge $+q$ end on a charge $-q$. Thus the so-called *induced charge* on G is equal and opposite to the inducing charge q.

At a point between the plates and not too near their edges, the field will, from symmetry, be directed down from B to G in Fig. 7–8. Since we are neglecting edge effects, we have no reason to assume the field to be different at one place than at another and so the field must be uniform.

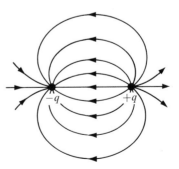

FIG. 7–9
Electric field due to a dipole.

Illustration 2. *Electric field due to a dipole*

An electric dipole is defined as two equal and opposite charges that are a relatively small distance apart. In the (b) part of the example given in

Section 7–4 the two charges constitute a dipole for an observer some distance away. Whereas in that example we computed the direction of **E** at only three points, Fig. 7–9 shows, with the aid of lines of force, the whole overall pattern of the field of a dipole and the relative strength of the field in different localities.

7|6 ELECTRIC POTENTIAL

We saw in mechanics that it was convenient to introduce the concept of potential energy in connection with gravitational and certain elastic forces. The work W that one must do against gravity in climbing a mountain is the same regardless of the path taken; it is just a question of one's gain in height, h. So we equate the work done to what we call the gain in potential energy, ΔV, and say that if one's PE at the bottom is V_A, then at the top it will be

$$V_B = V_A + \Delta V,$$

where $\Delta V = W$.

We can do the same thing in connection with electrostatic fields. The work needed to carry a charge from a point A to a point B in a field due to charges at rest does not depend on the path taken, so we can introduce the concept of electric potential energy. We shall make one difference, though, and talk about *work per unit charge*, rather than the work itself.

Consider a region in which the electric field has the magnitude E. If the field is uniform, as it is between parallel plates, E is, of course, the value of the field at all points. If the field varies, as it does in the neighborhood of a point charge, then we should consider a region small enough so that the percent variation in E is not large, and regard E as the average value of the field throughout this region. Let us place at a point A in the region a charge q', a positive value of q' meaning a positive charge, a negative value of q' meaning a negative charge. Here the prime is used just to remind us that q' is *not* a charge contributing to the field, but a charge on which the field due to other charges acts. If q' is positive the force on it will be $\mathbf{F}_e = q'\mathbf{E}$, in the direction of **E**. Let us push the charge a distance d in the opposite direction, to the point B, so that we will have to do the work

$$W_{AB} = F_e d = q'Ed \qquad\qquad \textbf{7-5}$$

on the charge. Moving the charge in the direction opposite to that in which the field urges it to go corresponds to lifting a stone against the force of gravity, or to going up a mountain, in which cases the potential energy increases by W_{AB}. Let V_{AB} represent the gain in electrical PE *per unit charge*. Then

$$V_{AB} = \frac{W_{AB}}{q'} = Ed, \qquad\qquad 7\text{-}6$$

where d is the displacement opposite to **E**. If the displacement is oblique to **E**, its projection in the direction opposite to that of **E** must be taken to be d in Eq. (7–6), which is frequently written in the form

$$E = V_{AB}/d \qquad\qquad 7\text{-}7$$

and used to compute the magnitude of **E** when V_{AB} and d are known.

Change in electrical PE per unit charge is called *potential difference* (abbreviated to pd). From Eq. (7–6) we see that in the mks-coulomb system the unit of pd is the joule per coulomb; since this unit is used frequently it has been given the familiar name *volt* (V), so that

$$1 \text{ volt} = 1 \text{ joule/coulomb} \qquad (1 \text{ V} = 1 \text{ J/C}). \qquad 7\text{-}8$$

From Eq. (7–7) it follows that **E** may be expressed in *volts per meter*, as well as in newtons/coulomb, hence

$$1 \text{ V/m} = 1 \text{ N/C}. \qquad\qquad 7\text{-}9$$

Proof of this equality is left as a problem.

While we have only defined the pd between points in a field, it is often convenient to choose some reference point (say A) and arbitrarily call its potential zero. Then the potential at B is said to be V_{AB} volts above zero. Such a zero reference point is usually taken to be the earth, since most terrestrial systems are electrically connected to the ground at some point.

Electrostatic potential difference or "voltage" may be measured with a gold-leaf electroscope (Fig. 7–10). The knob B is at the top of a metal rod and attached to the lower end of this rod is a strip of thin metal foil (gold or aluminum) which can swing out as shown. This system is protected from air currents by enclosing it in a box with glass windows. A charged conductor at an unknown voltage above or below that of the ground, such as the upper plate in Fig. 7–8, is connected to the knob B of the electroscope. The rod and leaf share charge of the same sign as that on the conductor. Since like charges repel and the leaf can move, the leaf swings out to where the electrostatic repulsion just keeps the leaf from falling under its own weight. The greater the pd between B and ground, the greater are the charges on rod and leaf, and the greater is the resulting deflection. While such an instrument must be calibrated by applying known potential differences, we shall see in Experiment 12 that a pd may be determined absolutely, that is, in terms of the definition $V_{AB} = W_{AB}/q'$.

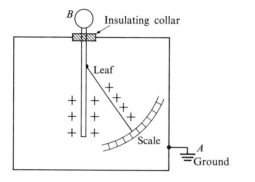

FIG. 7-10
Electrostatic voltmeter, or gold-leaf electroscope.

Illustration. *Lightning*

The strong vertical air currents in thunderstorms cause frictional electrification of the water drops in a thundercloud. Large potential differences develop between the top and bottom of the cloud and between the bottom of the cloud and ground. Now air is normally a poor conductor of electricity, but if it is subjected to an electric field greater than 3×10^6 V/m air becomes a very good conductor. Very strong electric fields literally tear electrons from the atoms that compose the molecules of air and lightning flashes occur. Of most interest are the discharges between the bottom of the cloud and the earth below.

The lower part of a thundercloud is usually negatively charged. This repels electrons from the tops of trees and buildings into the ground, so that these exposed objects become positively charged (Fig. 7–11) and are more attractive goals than the ground for lightning discharges to earth. For this reason one should never take shelter under a lone tree, or expose oneself in an open field, on water, or on mountain ridges during thunderstorms. Some warning may be provided by the building up of induced charge on one's hair, causing it to stand on end. Mountain climbers are aware of this and "the sound of the bees," a buzzing due to a continuous slow discharge from pine needles and sharp rocks, and wisely seek protection below a cliff or in a cave. Lightning rods, which were first proposed by Benjamin Franklin, do afford much protection to a building provided

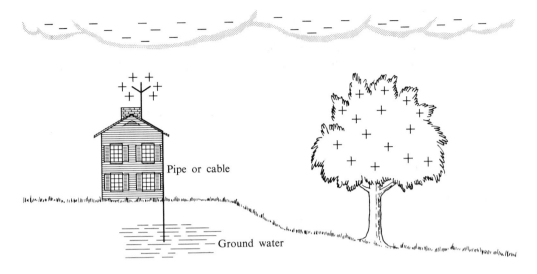

FIG. 7-11. Electrification produced by thundercloud and lightning protection.

they make very good connection with the earth, say through a copper pipe that runs down the corner of the building and then deep enough into the ground to reach moisture.

7|7 THE VAN DE GRAAFF ELECTROSTATIC GENERATOR

Early devices for producing high voltages and electrostatic discharges were friction machines with rotating glass plates. The most popular modern type of electrostatic generator is the Van de Graaff, in which a hollow conducting sphere is mounted on an insulating hollow cylinder containing a moving belt, as shown in Fig. 7-12. Charge is sprayed onto the belt at the bottom and taken off at the top. It is no problem to transfer charge on the belt to the *inside* of the hollow sphere, because there is no electric field there. The charges repel one another and push out to the outside surface of the sphere, raising its potential. Eventually a potential is reached at which charge leaks off (mainly into the air) as fast as more charge is brought up by the belt. This potential increases with the radius of the sphere; it may also be raised by enclosing the whole apparatus in a steel tank containing a suitable gas at high pressure, for which the breakdown limit of the electric field is far greater than 3×10^6 V/m. In this way potentials millions of volts above ground have been attained.

171

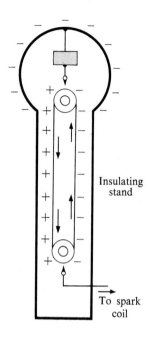

Insulating
stand

To spark
coil

Large potential differences may be used to give high energy and speed to charged atomic particles. If a particle of charge q is taken from A to B against an electric field, the work done on it is $W_{AB} = qV_{AB}$ (from the definition of pd). If the particle travels back from B to A the field will pull it along and accelerate it. Conservation of energy requires that the kinetic energy gained must equal the work done, so if the particle starts from B with practically no initial velocity it will reach A with a speed v given by

$$\text{KE gained} = W_{AB},$$
$$\tfrac{1}{2}mv^2 = qV_{AB},$$
$$v = \sqrt{2qV_{AB}/m}, \qquad \qquad \textbf{7-10}$$

where m is the mass of the particle. We shall refer to this relation in the next chapter.

FIG. 7-12. Van de Graaff generator.

7|8 THE COULOMB BALANCE AND DETERMINATION OF THE CONSTANT k_e

Parallel charged plates find many uses in electronics and atomic physics because they provide (except near the edges) a uniform and measurable elective field **E**. From Eq. (7-6) we have for the magnitude of **E**:

$$E = V_{AB}/d.$$

Such plates may also be used to determine experimentally the fundamental constant k_e in Coulomb's law. It is impractical to measure the force between two small charges with a known separation, but it is possible to measure the force of attraction between oppositely charged plates with a separation d, area A, and a pd $= V_{AB}$. Since this latter force is also an electrostatic attraction, F_e, the formula relating it to d, A, and V_{AB} may be derived from Coulomb's law and, as might be expected, this formula also contains the constant k_e that determines the strength of electric forces. A sketch of the apparatus is shown in Fig. 7-13.* The experimental procedure is as follows.

_____

* Now purchasable from the W. M. Welch Co., 7300 N. Linder Ave., Skokie, Ill. 60076.

Step 1. With a small weight of mass m (say 50 mg), but no voltage, on the upper suspended plate, the telescope is sighted on the lower edge of this plate.

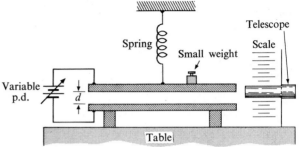

FIG. 7–13. Coulomb balance.

Step 2. The weight is gently removed. The spring contracts slightly and the upper plate rises. The pd V_{AB} between the plates is now gradually increased until the electric force F_e replaces the weight mg, in which case the bottom edge of the upper plate will again be lined up with the crosshairs of the telescope. Then $F_e = mg$. V_{AB} is recorded with a voltmeter and d measured. The area A of the plates may be found by measuring their diameter. The only unknown in the formula* for F_e is k_e and this may now be determined.

7|9 THE MILLIKAN OIL-DROP EXPERIMENT

We are now in a position to understand one of the great experiments of all time, one which firmly established the atomicity of electricity and the size of the elementary charge. This experiment was performed by the American physicist Robert A. Millikan (1868–1953), and for his work he later received the Nobel Prize. Millikan's apparatus is shown in Fig. 7–14. Two parallel horizontal metal plates are separated about 1 cm by insulators and shielded from air currents by an enclosing box. Oil is sprayed as fine droplets into the region between the plates. This region is brightly illuminated and viewed against a dark background through a telescope. The oil drops appear as tiny bright stars.

First a variable pd V_{AB} is applied to the plates. The oil drops are frictionally charged by the spraying process. Let q be the charge on a certain drop of mass m. The weight of the drop is mg downward. The electric

---

* Should the reader wish to use the formula (there is no need to commit it to memory), it is $F_e = A V_{AB}^2 / 8\pi k_e d^2$.

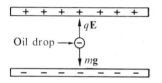

FIG. 7–14. Millikan's oil-drop experiment.

force on the drop $F_e = qE$, where E is the magnitude of the electric field due to V_{AB}. The direction and magnitude of this field may be adjusted so that the electrical force is upward and just balances the weight of the drop. Then for the suspended drop we have

$$qE = mg. \tag{7-11}$$

In this equation, E may be computed from the formula $E = V_{AB}/d$, by measuring V_{AB} (with a voltmeter) and the plate separation d. The mass of the drop must be determined from its size and density. The density of the oil may easily be found, but the drops are too small for one to measure their diameters directly. Millikan found that he could measure the radius of a drop very accurately by timing its rate of fall when the electric field was turned off. The drops very quickly reach a limiting velocity of fall which is determined by their size, their density, and the viscosity of the air. (The correct formula relating these quantities had been worked out by a theoretician named Stokes.) After finding mg for a drop, Millikan corrected for the buoyancy of the air.

Millikan's results showed that for every drop, q was a whole number multiple of the charge $\pm e$, where e is evidently the *fundamental quantity of charge out of which all larger charges are built*. His drops had charges such as $+4e$, $-5e$, $+2e$, but never charges like $+3.72e$, or $-0.64e$.

The measured value of e is $e = 1.60 \times 10^{-19}$ C.

1. A conducting body receives negative charge at the rate of 10^{-9} C/sec.

   a) About how many electrons does it receive per second?

   b) If the body initially has a positive charge of 10^{-8} C, what will its charge be 10 sec later? 20 sec later?

2. Two charged balls 4 cm apart repel each other with a force of 36×10^{-5} N. What will the force be if

   a) the same charges are placed on balls 3 cm apart, and

   b) the separation is 6 cm and each charge is doubled?

3. Two identical conducting balls 4 cm apart have charges of

$$+12 \times 10^{-9} \text{ C} \quad \text{and} \quad -4 \times 10^{-9} \text{ C},$$

respectively. The balls are touched together and again placed 4 cm apart; they now repel rather than attract each other.

   a) Why did the force change from attraction to repulsion?

   b) Find the final charge on each ball.

   c) Find the original force and the final force between the balls.

4. Find the electric field $\mathbf{E}$ at the point $x = 0$, $y = 0$ when a charge of $+64 \times 10^{-9}$ C is at $x = 0.4$ m, $y = 0$ and a charge of -27×10^{-9} C is at $x = 0$, $y = 0.3$ m.

5. Represent with lines of force the electric field due to two equal positive charges.

6. Show that if the electric field is horizontal and decreasing in the positive x-direction, then there must be a negative space charge in the region.

7. Parallel plates are separated by 8 mm. If the upper plate is charged to 40 V and the lower plate is grounded, find

   a) the field $\mathbf{E}$ between the plates, and

   b) the force on an electron located between the plates.

8. a) In problem 7 how much work is needed to move the electron from the positive to the negative plate?

   b) What kinetic energy will the electron gain in traveling from the negative to the positive plate?

   c) If in (b) the electron ($m = 9 \times 10^{-31}$ kg) starts from rest, with what speed will it strike the positive plate?

9. An *electron-volt* (eV) is a small unit of energy convenient for atomic physics; it is defined as the energy acquired by an electron that is accelerated through a pd of 1 V. Show that $1 \text{ eV} = 1.6 \times 10^{-19} \text{ J}$.

10. Prove that a volt per meter equals a newton per coulomb.

11. A set of parallel plates is charged to a potential difference of 100 V and then the charging source is disconnected. If the separation of the plates is now increased from 0.5 cm to 2 cm, what (neglecting edge effects) will be

a) the potential difference across the plates, and

b) the electric field between the plates?

c) Did the field change?

d) Did the electric energy increase and if so, where did the added energy come from?

12. Same as Problem 11 except that the charging source remains connected to the plates.

13. If a Van de Graaff generator carries a charge of 10^{-3} C/sec up onto the insulated sphere, how much work per second, or power, is required when the sphere has reached a potential of 2×10^6 V?

14. A hollow conductor bears a net charge q_1. A charge q_2 is placed on an insulated body inside the conductor. Find the charge on

a) the inside surface, and

b) the outside surface of the conductor.

Experiment 11   FIELDS

Object: To study, as typical fields, the magnetic fields of magnets and the earth.

Theory: The concept of a *field* was invented to describe (or explain) action at a distance. Thus since gravitational, electric, magnetic, and nuclear forces may exist between particles in a vacuum, we speak of gravitational fields, electric fields, etc.

Here we shall study magnetic fields because they are the easiest to set up and measure. However, since we shall be dealing with steady magnetic fields due to permanent magnets, there will be a complete analogy between our magnetic fields and certain electric fields. If you wish, you may call all north poles "positive charges" and all south poles "negative charges" and you will have the electrical analogue. Hence imagine lines of force to radiate out from north poles toward south poles. Remember that the intensity of the field due to a point charge falls off inversely with the square of the distance; this is true also for the magnetic field due to a magnetic pole. Since fields have direction, they must be added vectorially. A magnetized iron filing or a small magnetic compass should align itself with the resultant field. The compass is more sensitive, but iron filings give us quickly a picture of the whole field pattern in regions where the field is strong enough to magnetize the filings.

Procedure

Step 1. Using iron filings, form in turn the magnetic fields listed below and make a sketch of each. Place the magnets on the table, separated about 5 cm, place a sheet of cardboard over them, sift filings onto the cardboard and tap.

a) Magnets end to end, unlike poles adjacent. Investigate only the field between the magnets.

b) Magnets end to end, like poles adjacent.

c) Magnets in T-formation, pole of one about 5 cm from middle of other magnet.

Step 2

a) Fasten down a sheet of paper about 2 ft square. Using a compass, try to find a location where the earth's field is fairly uniform. Place a magnet near one end of your sheet and outline it on the paper, marking the poles. Indicate at one corner the general direction of the earth's field.

b) To trace a line of force, put a dot on the paper, set the case of the compass so that the dot is at the edge of the case directly in line with the needle, make a dot close to the case at the other end of the needle. Make use of the second dot to locate a third, and so on.

c) Enough lines must be traced to show the form of the field over the entire paper. Trace a line until it either runs off the paper or into the magnet, but also try to let the spacing of your lines be in accordance with the *strength* of the field as indicated by how violently the compass oscillates when shaken. Put arrows on each line to show the *direction* of the field.

Step 3

a) Look for a region where the field has almost zero intensity. Try to explain this vectorially.

b) Check by vector addition the direction of the field at two other points. *Question:* What is the polarity of the earth's equivalent magnetic pole up in the arctic?

Experiment 12 ELECTRICAL HEATING AND THE CALIBRATION OF A VOLTMETER

Object: To determine an unknown potential difference by measuring the work done when a known charge passes across this pd and to check a voltmeter recording this pd.

Theory: As explained in the next chapter, a continuous flow of charge constitutes an electric current and a flow of one coulomb per second is called an *ampere* (A). Electric currents are measured with *ammeters* and these may be calibrated directly in terms of the definition of the ampere, as in Experiment 13. If a current of I A ($=$ C/sec) passes for t sec through a coil of wire across which the pd is V V ($=$ J/C), then a charge $q = It$ coulombs has passed across the pd of V joules per coulomb and the work in joules done on this charge must be

$$W = qV = ItV. \tag{1}$$

We shall measure W by letting this work go to heat the equivalent of m kg of water ΔTK°. Experiments on heating water by doing a known amount of work through stirring have shown that 4185 J are required to raise 1 kg of water 1K°, but this same amount of energy will raise 11 kg of copper or brass, or 5 kg of aluminum 1K°. Since one cannot heat water without also heating its containing vessel, you should use a brass or aluminum cup that has been weighed and simply regard one-eleventh of its mass for brass (one-fifth for aluminum) to be its calorimetric "water equivalent," to be added to the actual mass of water in determining m. Then

$$W = 4185m \, \Delta T. \tag{2}$$

Procedure

Step 1. Using an arrangement similar to that shown in Fig. 7–15, heat a known mass of water from about 10° below to about 10° above room temperature (this will equalize heat exchanges between the cup and room) by passing a current I through a coil placed in the water.* Stir gently but steadily and note the time t that is required and your actual value of ΔT.

---

* A ΔT of, say, 20° centigrade is the same as a ΔT of 20° Kelvin. Why?

Read the current with an ammeter in series and note the reading of a voltmeter placed in parallel with the coil. Use current and voltage that best employ the full ranges of the meters available. (Why?)

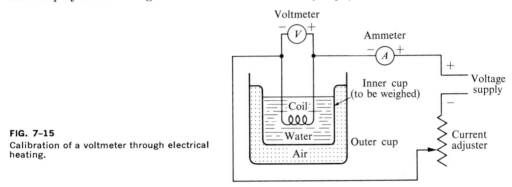

FIG. 7–15
Calibration of a voltmeter through electrical heating.

Step 2. Compute W from Eq. (2) and then V from Eq. (1). Compare your value for V with the voltmeter reading. Can you account for the percent difference in terms of the possible percent errors in m, ΔT, I, t, and the reading of V? (Meters are not reliable to closer than one scale division.)

8|1 THE ELECTRIC CURRENT

A succession of moving charges constitutes an electric current. The most common example of such a current is that due to the flow of electrons in a wire. In such a case we define the *electric current* I at a given point in the wire as the time rate of passage of charge through a section of the wire at that point. Let q be the charge passing in the time interval t. Then, by definition,

$$I = \frac{q}{t} \cdot \qquad \textbf{8-1}$$

It is conventional to take the positive direction of the current to be opposite to the way the electrons flow; as has been said, this convention results from Franklin's choice (long

Ampère's Law of Magnetic Force

before the discovery of the electron) of positive for the charge left on glass rubbed with silk. In electrolytes and ionized gases an electric current may arise from the combined motions of positive charges in what we call the direction of I and of negative charges moving in the opposite direction.

The mks-coulomb unit of current, called the *ampere* (or A) is defined as

$$1 \text{ ampere } = 1 \text{ C/sec.} \qquad \textbf{8-2}$$

The electric current at a point may be steady or variable with time. In the latter case, Eq. (8–1) defines the *average* current.

Currents usually flow in completely closed circuits with a constant value of I around the circuit; such a circuit is like an endless rotating belt. When a charged particle spins about its own axis, as most elementary particles are believed to do, each element of its charge constitutes a small circular current circuit.

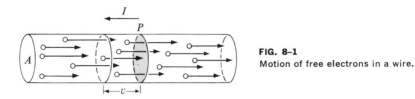

FIG. 8-1
Motion of free electrons in a wire.

Suppose that we have a wire of cross section A in which there are n electrons of charge $-e$ that are free to move per unit volume. Let v be the average drift speed of the electrons along the wire and I the resulting current. If we take a section through the wire at some point P (Fig. 8–1), we may say that all of the moving electrons that are drifting toward this section and are not more than a distance v from it will, on the average, cross this section in unit time. Measuring back a distance v along the wire, we see that there are nvA electrons in this length v of the wire, and they will cross our section in unit time, so that charge of magnitude $q = nvAe$ passes P in the time interval $t = 1$. Hence the current I (in the direction opposite to that in which the electrons are drifting) will be given by

$$I = \frac{q}{t} = nvAe. \qquad\qquad 8\text{-}3$$

Example 1. Suppose that people line up before a theater in a neat queue 8 ft wide, that there is one person to every 4 ft^2 of pavement covered by the queue, and that the line advances 12 ft/min. How many people enter the theater per minute?

Solution.   Any person within 12 ft of the door of the theater will enter in the following minute so that 12 ft $\times$ 8 ft $= 96$ ft^2 of the queue will reach the door per minute. With one person to every 4 ft^2, we see that $\frac{96}{4} = 24$ persons will enter per minute.

Example 2. A wire whose cross section is 1 mm^2 carries a current of 1 A. Assume that $n = 8.0 \times 10^{28}$ free electrons/m^3, a reasonable value for copper. Find the average drift speed of the free electrons.

Solution.   Here $I = 1$ A, $A = 10^{-6}$ m^2, $n = 8.0 \times 10^{28}$/m^3, $e = 1.6 \times 10^{-19}$ C. From Eq. (8–3) we find for v, the magnitude of the average drift velocity,

$$v = \frac{I}{nAe} = \frac{1 \text{ C/sec}}{(8 \times 10^{28}/\text{m}^3) \times (10^{-6} \text{ m}^2) \times (1.6 \times 10^{-19} \text{ C})}$$

$$= 7.8 \times 10^{-5} \text{ m/sec}.$$

Note the very low value of the drift velocity. Free electrons in a conducting wire do *not* move with anything like the speed of light, or with the speed of atomic particles. Frequent collisions with the atoms of the wire prevent the free electrons from picking up much speed down the wire. It is the tremendous number of free electrons per unit volume that accounts for the large currents of an ampere or more that are frequently sent through electrical lines. From the electrostatic point of view, an ampere *is* a large current because it involves the passage through a section of the wire of a coulomb every second, whereas a static charge of 10^{-6} C is relatively large. However, it should be realized that an electric current in a wire does not involve *separation* of positive and negative charges, whereas static electrification does. In a wire the free electrons simply move *between* the positive ions formed by the atoms that have contributed these free electrons to the wire. Such a wire does not bear a net charge.

8|2 THE MAGNETIC FORCE

The earliest recorded magnetic phenomena are those that the Greeks discovered in connection with the ore named *magnetite*, so called because it was found near the city of Magnesia in Thessaly. Magnetite is now known to be iron oxide (Fe_3O_4). The Greeks, and later the Chinese, found that pieces of magnetite attract to themselves unmagnetized iron, the attraction being most pronounced at certain regions called *poles*. It was observed that in this way pieces of iron could be permanently magnetized, and that when a rod of magnetized iron was suspended about a vertical axis, it would align itself in a north-south direction. The north-seeking pole was called the *north pole*, the other the *south pole*. The ancients were thus led to use the magnetic compass in navigation.

Later the forces between magnets were observed and studied. Such *magnetic* forces, which occur between uncharged magnets, are obviously different from electrostatic forces. Coulomb made a careful quantitative study of the forces between magnetic poles and arrived at another general law, very similar to his law of electrostatic force, which states the following.

Like poles repel, unlike poles attract, the force being proportional to the product of the pole strengths and inversely proportional to the square of their distance of separation.

This relation was used to define a unit of pole strength.

For over a century this law of Coulomb was taken to be one of our fundamental laws and pole strength was considered a fundamental concept. Now, however, a more unified way of describing the electric and magnetic properties of the physical world has been adopted and this way will be described.

Volta's discovery, in 1800, in Italy, that an electric current will flow continuously in a wire joining strips of, say, copper and zinc dipped into an acid or salt solution, opened up the possibility of studying the properties of electric currents. One such important property was discovered in 1819 by the Danish scientist Oersted, who observed that a pivoted magnet was deflected when in the neighborhood of a current-carrying wire (see Fig. 8-2). The importance of this discovery lies in the fact that it showed for the first time an interrelation between electricity and magnetism; it showed that a force can exist between electric currents and magnetic poles. In present-day theory this force is regarded as being of the same fundamental nature as that which exists between neighboring poles, that is, a *magnetic* force.

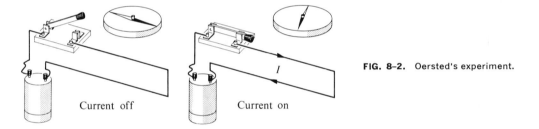

I

FIG. 8-2. Oersted's experiment.

Current off   Current on

If a magnetic force exists between a magnetic pole and an electric current, is it not a logical step to look for a magnetic force between neighboring currents? Such an effect was soon found, and a careful experimental study of it led the French physicist Ampère (1775–1836) to the statement of the general expression for the magnetic force between two current circuits; this expression, in turn, is based on one representing the magnetic force between the moving charges that constitute such circuits.

The final step in the development of our present-day theory of magnetic forces has been to drop the concept of magnetic poles as separate entities, comparable with electric charges. On the one hand, positive and negative charges *can* be completely separated, and there are atomic particles, such as electrons and protons, that have a net charge of one sign or the other. There are conducting media in which charged particles can move and so give rise to an electric current. On the other hand, magnetic north and south poles always occur in pairs and *cannot* be completely separated and isolated. Extensive search has not revealed any magnetic equivalent of the electron and there are no magnetic conductors in which poles can move and produce magnetic currents. It is possible to show that the magnetic forces which magnets exert on one another may be attributed to magnetic forces between charges moving within the magnetized bodies. The force between an electric current and a magnetic compass is similarly attributed to the forces between the charges moving within the magnet and the moving charges that constitute the electric current. Thus we postulate that all magnetic effects are electrical in origin and are due, not to electrostatic forces and fields, but to those resulting from the *motion* of charges.

8|3 AMPERE'S LAW OF MAGNETIC FORCE. FUNDAMENTAL LAW IX

Ampère was interested in describing the magnetic force between neighboring current circuits. He found experimentally that this force depends not only on the electric current in each circuit, but also on the size, shape, and relative positions and orientations of the two circuits. The general problem seemed hopelessly complex, so Ampère and his contemporaries (notably Biot and Savart) turned to the special case of two long straight wires carrying currents I and I', respectively (see Fig. 8–3). They found that the magnetic force is attractive when the currents are in the same direction and repulsive when the currents are opposite, and that the magnitude of the force is proportional to the product of the currents.

FIG. 8–3
Magnetic force between long parallel current circuits.

When the separation of two parallel current circuits is much less than their length, the force is found to be inversely proportional to the separation, but this is just the relationship that would result if F_m, the magnitude of the magnetic force between neighboring moving charges, were an *inverse square force*. The proof of this statement, which involves the calculus, may be found in more advanced texts. However, the reader should note that the force F_m on a segment of the lower current-carrying wire in Fig. 8–3 is due not just to the charges moving in that part of the upper wire immediately above the segment; *all* the charges moving in the upper wire, from far to the left to far to the right, exert a force on the charges moving in a given segment of the lower wire.

185

To arrive at the correct expression for the magnetic force between any two given current circuits, one must postulate that the magnetic force $\mathbf{F}_m$ between two moving charges has the following properties:

1. F_m is proportional to the product of the charges.

2. F_m is proportional to the product of their velocities.

3. F_m is inversely proportional to the square of their distance of separation.

4. $\mathbf{F}_m$ is dependent in both magnitude and direction on the directions in which the charges are moving.

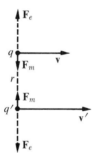

FIG. 8–4. Magnetic force between moving charges.

Thus the magnetic force cannot be described as simply as can the electrostatic force between two charges. The closest resemblance between the expressions for F_m and F_e occurs in the special case of two charges, q and q', that are moving side by side in parallel directions with speeds v and v', respectively (Fig. 8–4). Call the distance between the charges r. Then the magnitude of the magnetic force, which acts in addition to the coulomb force, is given by

IX. $$F_m = k_m \frac{(qv)(q'v')}{r^2} ,$$ 8–4

where k_m is a constant of proportionality related to the size of the unit we choose for measuring charge. The *coulomb* (C) *is defined* as that unit of charge for which

$$k_m = 10^{-7} \text{ N-sec}^2/\text{C}^2$$ 8–5

when q and q' are expressed in terms of this unit, v and v' are in meters per second, r is in meters, and F_m is in newtons. If velocities to the right are taken as positive and ones to the left as negative, then a positive value for F_m means that the two charges experience a magnetic attraction, while a negative F_m means a repulsion. Equation (8-4) is a special case of Ampère's law for magnetic force, our *Fundamental Law IX*. From this law one may deduce the expression for the magnetic force between long parallel current circuits, namely,

$$\frac{F_m}{L} = \frac{2k_m I I'}{d},$$
8-6

where F_m/L is the magnetic force per unit length, I and I' are the respective currents, d is the separation of the wires, and k_m is the same constant as in Eq. (8-4).

Equation (8-6) may be tested experimentally (see Experiment 13 at the end of the chapter). One may show that F_m is proportional to the product of the currents and inversely proportional to the separation d. One may also use this experiment to calibrate instruments, called *ammeters*, that are used to measure electric currents. To do this one makes $I' = I$ and measures F_m, L, and d. Then, choosing $k_m = 10^{-7}$ N-sec^2/C^2, one calculates I^2 and hence I. With the above choice for k_m the value of I will, by definition, be in coulombs per second. We call

1 C/sec = 1 ampere (A),

so I comes out in amperes. An ammeter in the circuit should (within experimental errors) read this number of amperes.

Of course Ampère's general law may be applied to deduce other expressions representing the magnetic force between current circuits of other shapes and sizes and these expressions may also be tested experimentally. In this way a firm experimental foundation for Ampère's law has been established.

8|4 THE RELATIONSHIP BETWEEN k_m AND k_e

While the value of k_m may be chosen quite arbitrarily, some judgment should be exercised in its choice. In the past physicists have been tempted to put $k_m = 1$, but in so doing they forced upon themselves a set of *electromagnetic units* which do not coincide with the practical units. Fortunately the electrical engineer's units (ampere, volt, ohm, watt, etc.) were so chosen that by taking 10^{-7} as the value for k_m and using mks mechanical units, all of our electrical units become the same as the *practical or electrical engineering units*. Since the value chosen for k_m determines the size of the coulomb, when k_e is measured, using a Coulomb balance, the value found experimentally for k_e also depends on our choice for k_m. However, it is an interesting fact that the *ratio k_e/k_m does not depend on the value chosen for k_m*, so that it is this ratio which is a fundamental constant of our physical world. Let us see what the value of this ratio is.

When k_m is taken to be 10^{-7} N-sec^2/C^2, the best experimental value for k_e is found to be close to 9×10^9 N-m^2/C^2. Thus we find that

$$\frac{k_e}{k_m} = \frac{9 \times 10^9 \text{ N-m}^2/\text{C}^2}{10^{-7} \text{ N-sec}^2/\text{C}^2} = 9 \times 10^{16} \text{ m}^2/\text{sec}^2,$$

which coincides, with regard to both magnitude and units, with the square of the experimental value for the speed of light in air or empty space. Thus

$$\sqrt{k_e/k_m} = c = 3 \times 10^8 \text{ m/sec}, \qquad \textbf{8-7}$$

where c represents the speed of light *in vacuo*.

Is this a coincidence or is there a connection between electric and magnetic forces and light? This question was answered by Maxwell, and his theory will be discussed in the next chapter. However, we may say here that physicists always suspect that coincidences such as the above have an underlying significance and are not the result of mere chance.

Electric and magnetic forces are certainly related. Equation (8–7) is one indication that this is so and another is that electric and magnetic forces are both forces between *charges*. Now let us do a little "thought experiment." Imagine two charges, q and q', moving eastward side by side, each with the speed v relative to us on the ground. These charges might be electrons moving in parallel wires, in which case they would move between the positive ions. Assuming that the wires bear no electrostatic charges, we would attribute the observed force between the wires to the magnetic force between pairs of charges such as q and q'. Next picture another observer who is moving eastward, also with the speed v. The force between the current-carrying wires exists, independent of the observer, but how does the moving observer interpret it? He could say that while the electrons in the wires appear to be at rest relative to him, positive ions in equal numbers seem to be moving westward with the speed v and that they account for the magnetic force between the wires. Finally, suppose that our two charges are not in wires but are moving by themselves in an evacuated tube. We would describe the force between the charges as the resultant of their electrostatic and magnetic interactions, yet to our moving observer the charges would appear at rest and hence subject to an electrostatic force only. This is leading us into the theory of relativity, which deals with problems concerning observers in relative motion (see Chapter 10). We need not delve into the intricate reasoning of relativity to see that what we call electric and what we call magnetic forces are dependent on our point of view and that these two forces are two aspects of what is called the *electromagnetic force*. This is our second fundamental force in nature, gravity being the first from an historical point of view. We have seen that this second force is many times stronger than gravity and that it is the force which bodies experience when they possess that property called "charge."

8|5 THE MAGNETIC FIELD

We found it convenient to describe electrostatic forces in two steps by saying that an electric field **E** exists in the neighborhood of a charge and that a second charge q' placed in the field will experience the force $\mathbf{F}_e = q'\mathbf{E}$. Since the magnetic interaction is much more complicated, it becomes all the more advisable to treat it in two steps also. So we say that a magnetic field **B** exists around a moving charge and that a second charge q' moving in this field will experience a force $\mathbf{F}_m$. We need not trouble ourselves with the general expressions for the field and for the force—they are rather complicated—but we shall take a simple yet important case by way of illustration.

Let us look again at Fig. 8-4. The magnetic force between the two moving charges is given by Eq. (8-4) and the statement that a positive F_m means attraction, a negative F_m repulsion. The two-step field way of saying the same thing is as follows:

a) Around the moving charge q there exists a magnetic field **B**, which is a vector quantity.

b) The *direction* of the magnetic field is in circles about the line of motion of the charge. The fingers of one's *right* fist will curl in the positive sense of the field when the thumb of the right hand points in the direction of motion of a positive charge or the direction of a positive current, as in Fig. 8-5. (Treat negative charges as positive charges moving the other way.) In Fig. 8-4, **B** due to q is into the paper at q'.

c) The *strength* of the magnetic field at the position of q' is

$$B = k_m qv/r^2, \qquad\qquad \textbf{8-8}$$

r being the distance from q to q'. This completes the step describing the field.

d) The *force* of a magnetic field **B** on the charge q' moving with speed v' has the magnitude

$$F_m = q'v'B, \qquad\qquad 8-9$$

and its *direction* is at right angles to **v**' toward the region where the magnetic fields of q and q' are opposite to each other. (Note that for positive q and q' in Fig. 8–4 the fields are opposite between the charges and so q' is urged toward q.) This completes the step describing the force exerted by a magnetic field on a moving charge.

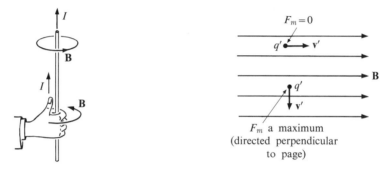

FIG. 8–5. Right-hand fist rule for finding the direction of the magnetic field **B** around a current-carrying wire.

FIG. 8–6. Dependence of magnetic force on direction of motion of charge.

We may use Eq. (8–9) to define the mks-coulomb unit for **B**. We have

$$B = \frac{F_m}{q'v'} = \frac{\text{N}}{\text{C-m/sec}}.$$

Since 1 C/sec is 1 A, we may also express **B** in terms of newtons per ampere-meter. So, if we take the product BIL, that is, magnetic field times current times length, we will come out in the mks-coulomb system with something measured in newtons, which means a force. This suggests—and a rigorous proof can be given—that BIL expresses the force of a magnetic field **B** on a wire of length L carrying a current I perpendicular to the field. Calling this force F_m, we have

$$F_m = BIL. \qquad\qquad 8-10$$

Experience shows that the magnetic force decreases to zero as the current-carrying wire is turned until parallel or antiparallel with the magnetic field **B**. This same direction dependence also applies to the magnetic force on a moving charge (Fig. 8–6).

Faraday introduced the concept of magnetic lines of flux, analogous to his electric lines of force discussed in the last chapter. It is customary to take a scale such that there are **B** lines per square meter through a surface perpendicular to **B**; that is, B determines the *concentration of the flux. The weber is the mks-coulomb unit of flux*, which will be represented by the symbol Φ.

Since magnetic fields are directed in circles about moving charges, lines representing such fields form closed curves and do not end. In the case of a bar magnet, lines representing **B** run from the north pole to the south pole outside the magnet and then from the south pole back to the north pole inside the magnet. Due to this difference between magnetic and electrostatic fields, and also due to the fact that a magnetic field always acts on a moving charge in a direction perpendicular to the motion, so that the field does no work, we cannot introduce a magnetic analogue of the electrostatic potential or pd.

Illustration 1. *A circular current circuit*

It is useful to be able to produce a magnetic field whose value may be computed from its definition, just as we may compute the electric field between parallel plates from their pd and separation. As an example of how this may be done we take the simple case of a current I flowing in a circular wire of radius R and compute **B** at its center C (Fig. 8–7).

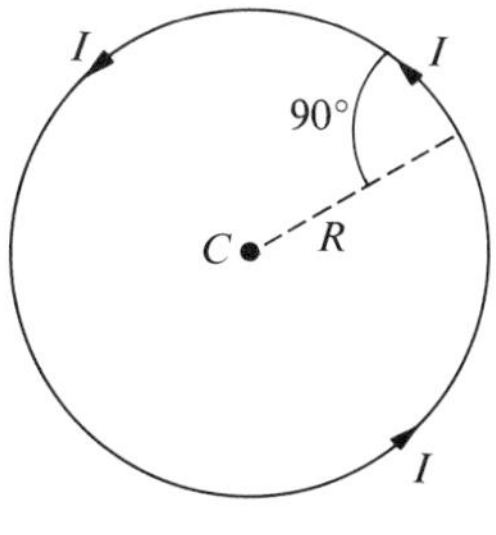

FIG. 8–7. Circular current circuit.

Call the total moving charge in the wire q and let v represent its average speed. The current I is equal to the charge passing any point in the wire in one second. The circumference is $2\pi R$, so that q circulates around the wire $v/2\pi R$ times per second. Hence

$$I = qv/2\pi R, \quad \text{or} \quad qv = 2\pi RI.$$

From Eq. (8–8) we have

$$B = k_m qv/R^2.$$

We substitute $2\pi RI$ for qv in this last equation and, after canceling an R in the numerator and denominator, get

$$B = 2\pi k_m I / R. \qquad\qquad \text{8-11*}$$

If we put $k_m = 10^{-7}$, express I in amps and R in meters, B will come out in newtons per ampere-meter. The direction of **B** at C for a counterclockwise current is, according to the right-hand fist rule, out from the paper.

A similar procedure may be applied to compute **B** in more complicated situations.

Let us now turn to the second part of Ampère's law, that is, the part dealing with the force exerted on a moving charge or current in a magnetic field.

Illustration 2. *A rectangular current-carrying coil in a uniform magnetic field*

Consider a rectangular coil of one turn suspended so that its plane is parallel to the magnetic field **B** (see Fig. 8–8). Let the length of the coil perpendicular to the field be l and its width w, and let the current in the coil be I. Such a coil is found in *galvanometers*, which are devices for measuring small currents, and in ammeters (which measure larger currents), and voltmeters (which measure potential differences).

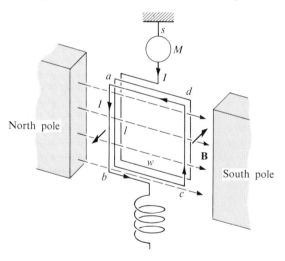

North pole

South pole

FIG. 8–8

Torque on a galvanometer coil. Force on vertical sides is BIl per turn.

---

* There is no need to commit this formula to memory.

The force $\mathbf{F}_m$ on the top (ad) and bottom (bc) of the coil is zero, since for these two sides the current I' is parallel or antiparallel to $\mathbf{B}$. For the side ab, which is perpendicular to $\mathbf{B}$, the force $\mathbf{F}_m$ will be BIl normal to the plane of the coil. On the opposite side cd, the force will have the same magnitude, but the opposite direction. The result is a twisting action and the greater the current the more this action will twist the suspension S and rotate the viewing mirror M.

*8|6 THE ACTION OF ELECTRIC AND MAGNETIC FIELDS ON CHARGED PARTICLES

Electric and magnetic fields have been and are being used to accelerate and deflect elementary particles and atomic ions. Such experiments usually enable one to determine (a) the speed v, and (b) the charge-to-mass ratio q/m, of the particle in terms of the E- and B-values of the applied fields.

The force $\mathbf{F}_e$ due to an electric field $\mathbf{E}$ acting on a particle of charge q is, as we have seen,

$$\mathbf{F}_e = q\mathbf{E},$$

the force on a positive particle being in the direction of $\mathbf{E}$ and that on a negative particle opposite to $\mathbf{E}$. The force $\mathbf{F}_m$ due to a magnetic field acting on a particle of charge q moving with the speed v has the magnitude

$$F_m = Bqv$$

when the charge moves perpendicular to $\mathbf{B}$, decreasing to zero when it moves parallel to $\mathbf{B}$. The direction of $\mathbf{F}_m$ is always perpendicular to the direction of $\mathbf{B}$ and that of $\mathbf{v}$, (see Fig. 8–6). Primes have been dropped since we are concerned only with one charge, namely that acted on by the field, and not with the charge(s) producing the field.

It is a good thing to remember in connection with experiments involving charged particles and fields that (1) gravitational forces are negligible, (2) an electric field $\mathbf{E}$ in the direction of motion will speed up or slow down the particle, depending on the sign of its charge, (3) an electric field $\mathbf{E}$ transverse to the motion will deflect the particle, (4) a magnetic field $\mathbf{B}$ parallel to $\mathbf{v}$ will have *no effect* at all, and (5) a magnetic field $\mathbf{B}$ transverse to the motion will deflect the particle, but not alter its speed.

---

* Sections 8–6 and 8–7 may be omitted without loss of continuity.

a | Determination of v

The three most commonly used methods of determining the speed of a particle are the following.

Method 1.   Acceleration of a charged particle through a known pd without collisions

In a vacuum tube, electrons that, in a sense, "boil off" from a hot filament may be accelerated toward a positive grid or tunnel-shaped electrode. Some of the electrons will strike the electrode and be captured, but others will shoot through the holes of the grid, or the tunnel, into the region beyond (Fig. 8–9). Let the potential of the grid (or tunnel) be V volts above that of the filament, so that the work done by the electric field on the electron (of charge $-e$) is eV. If the electron starts practically from a state of rest and does not suffer any collisions, we may equate the work done on it to its gain in KE, so that we have (see end of Section 7–7)

$$eV = \tfrac{1}{2}mv^2,$$
$$v = \sqrt{2eV/m}, \tag{8–12}$$

where v is the final speed and m the mass of the electron.

This same method may be applied to other particles of charge q (not necessarily e), provided that one knows that they are accelerated practically from rest.

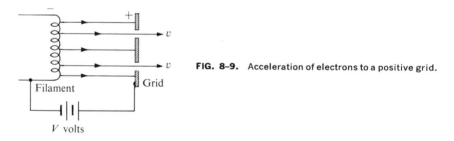

FIG. 8–9. Acceleration of electrons to a positive grid.

Method 2.   Balancing a moving charged particle in crossed electric and magnetic fields

Let a particle of charge q and mass m move horizontally into a region where an electric field $\mathbf{E}$ exerts an upward force $\mathbf{F}_e$ and a magnetic field $\mathbf{B}$ simultaneously exerts a downward force $\mathbf{F}_m$ on the particle. Here $\mathbf{v}$, $\mathbf{E}$, and $\mathbf{B}$ are mutually perpendicular, as in Fig. 8–10. If $\mathbf{F}_e$ and $\mathbf{F}_m$ are equal in magnitude, as well as opposite in direction, the net force on the particle will be zero and it will pass undeflected through the region of the fields.

FIG. 8-10
Crossed electric and magnetic fields.

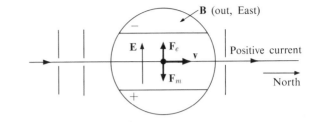

The necessary condition for this to happen is that

$$F_e = F_m, \qquad qE = qvB,$$
$$v = E/B. \qquad\qquad\qquad \textbf{8-13}$$

Method 3. Deflection of a moving charged particle in a transverse magnetic field only

Let the particle of charge q, mass m, and velocity **v** move at right angles to a uniform magnetic field **B**. Since $\mathbf{F}_m$ is normal to **v**, the latter will change in direction, but not in magnitude. In other words, the particle will undergo centripetal acceleration and move in a circular path whose radius we shall call R. $\mathbf{F}_m$ now provides the centripetal force $\mathbf{F}_c$ required for this circular motion. We therefore put

$$F_c = F_m, \qquad \frac{mv^2}{R} = qvB,$$
$$v = \frac{qBR}{m}. \qquad\qquad\qquad \textbf{8-14}$$

The paths of charged particles may be made visible in several ways; the paths may be photographed in cloud and bubble chambers.* In this way R may be measured. A known magnetic field may be generated by sending a known current through one or more circular coils and then computing B with the aid of a formula derived from Ampère's law, such as Eq. (8-11). If the nature of the particle is known, then its q/m ratio may be taken from tables and v computed from Eq. (8-14).

b | Determination of q/m

Suppose now that the q/m ratio of a certain particle has not yet been determined and we wish to measure its value. How would we proceed?

Many ingenious methods of measuring q/m for particles have been devised. The simplest ones involve a combination of two of the three

---

* See texts on modern physics.

methods just described for finding v, so that between two of the equations

$$v = \sqrt{2qV/m}, \qquad v = E/B, \qquad v = (q/m)BR,$$

we may eliminate v and solve for q/m. The charge q is usually known; it is most often one or two electron units (1.6×10^{-19} C) so that the determination of q/m leads to that of m. Such devices for determining m are called *mass spectrographs*.

8|7 THE CYCLOTRON AND ITS SUCCESSORS

The cyclotron was invented by the American physicist E. O. Lawrence (1901–1958). It is a device for imparting a very high energy W to a charged particle by giving it many small increments in energy, such as 100 increments each equal to $W/100$. By this method the high potential differences and resulting insulation problems of Van de Graaff and trans-former-rectifier devices are avoided.

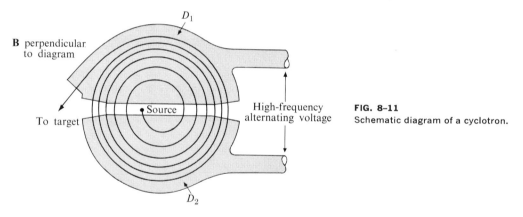

FIG. 8–11

Schematic diagram of a cyclotron.

The cyclotron (see Fig. 8–11) employs a magnetic field to bend the charged particles in a circular path inside an evacuated "tank." Within the tank are two hollow D-shaped electrodes which are connected to a source of alternating potential of high frequency and fairly high voltage. There is thus an alternating electric field **E** across the gap between the "dees." Suppose that a particle is accelerated across the gap and then, after completing a semicircular path under the influence of the applied magnetic field **B**, it again reaches the gap just when the electric field has been reversed. Such a particle will receive a second accelerating boost from the electric field. Then (and here is the important point) the particle will continue to be boosted every successive time it crosses the gap, so long as its mass remains constant and **B** does not change. The proof of this statement follows.

We have seen that for charged particles moving across a magnetic field **B** the radius of curvature R of their path must be given by Eq. (8–14), or

$R = vm/qB$.

The time t required for a half-revolution in a circular orbit of radius R will be

$$t = \frac{\pi R}{v} = \frac{\pi m}{qB} , \qquad\qquad \textbf{8-15}$$

a time that is not dependent on the speed v or radius R. (In a larger orbit the particle just goes proportionately faster.)

The particles spiral outward until they near the outer wall where, having attained their maximum energy, they are deflected toward some target.

We shall see in Chapter 10 that at speeds of over one-tenth the speed of light the mass of a particle is *not* constant, but increases. This increases the time t in Eq. (8–15), so that the particle arrives at the gap between the dees *after* the accelerating electric field has passed its maximum value. To compensate for this effect, one may (1) allow **B** to increase with m, or (2) allow the frequency of the alternating field to decrease with $1/m$ for a pulse of accelerating particles; after such a pulse has reached the target, the process may be repeated. Such devices are called *synchrocyclotrons*, since accelerations are synchronized with the relativistic mass increase.

To give heavy particles still higher energies, physicists have been forced to build larger and larger accelerators, since R increases first with v and then with m. It is much too expensive to build a cyclotron-type magnet with solid cylindrical poles many feet in diameter (the largest is 15 ft), and so in the large *synchrotrons or bevatrons*, which impart billions of electron-volts (BeV) of energy to particles, it has been necessary to use a ring of magnets to supply the deflecting magnetic field. If particles are preaccelerated up to speeds close to that of light, their speed will not undergo much further change, and so if the particles are held in a fixed orbit, the time for a revolution will also remain nearly constant. Then the magnetic field need be applied only at this orbit and not throughout its entire area. The synchrotron at the Brookhaven National Laboratories on Long Island, New York, is 840 ft in diameter and accelerates protons up to energies of 30 billion eV. Still larger accelerators are being built. The Atomic Energy Commission is planning a 200-BeV accelerator 4500 ft in diameter. An accelerator of a different type, favored by a number of laboratories because of its compactness, is shown in Fig. 8–12. **197**

FIG. 8-12. The 20 MeV "Emperor" accelerator at Yale University. For scale note the man in the center of the picture. The large pressure tank contains two Van de Graaff accelerators in tandem. Ions are accelerated linearly into the narrow tube; at the bend in the left foreground they are deflected by a magnetic field and directed to a target at the right. The momentum of the ions is measured by the strength of the magnetic field required for the 90° deflection. (Courtesy of Nuclear Structure Laboratory, Yale University)

Example. Protons are accelerated to a speed $v = 0.1c = 3 \times 10^7$ m/sec. Find the radius of curvature of their path in a magnetic field

$$B = 0.8 \text{ N-sec/C-m} = 0.8 \text{ N/A-m}.$$

Solution. In Eq. (8–14) we have $m = 1.6 \times 10^{-27}$ kg and $q = 1.6 \times 10^{-19}$ C for protons, so that

$$R = vm/qB$$
$$= \frac{(3 \times 10^7 \text{ m/sec}) \times (1.6 \times 10^{-27} \text{ kg})}{(1.6 \times 10^{-19} \text{ C}) \times (0.8 \text{ N-sec/C-m})}$$
$$= 3.75 \times 10^{-1} \text{ m}$$
$$= 37.5 \text{ cm}.$$

PROBLEMS

1. What is the magnitude and direction of the current due to a discharge through air in which 6×10^{17} electrons of charge $-e$ pass downward and 2×10^{17} ions of charge $+e$ pass upward through a given area in 0.1 sec?

2. The belt of a Van de Graaff generator is 40 cm wide and travels at a speed of 20 m/sec. If 5×10^{-6} C of charge is sprayed on every square meter of the outside face of the upgoing belt and -5×10^{-6} C/m^2 on the outside of the downgoing belt, what is the current (useful plus leakage) that will flow from the upper plate to ground?

3. Show that since like currents attract, the direction of the magnetic force on a charged particle moving perpendicular to an applied field **B** is toward the region where the magnetic field of the particle is opposite to **B**.

4. Show that the path of a charged particle moving obliquely to a magnetic field **B** is a helix, or spiral, about a magnetic line of force.

5. If charged particles from outer space, such as (1) electrons from solar disturbances, or (2) primary cosmic ray particles, enter the earth's magnetic field, to what areas of the earth's surface will such particles arrive in the

    a) greatest number,

    b) least number?

    c) Explain the relationship of this to the occurrence of Northern Lights.

6. Two parallel current circuits, each 30 cm long and 1 cm apart, carry the same current I. Compute the force between the circuits, both in newtons and in gram-weight, when $I = 5, 10, 15$, and 20 A, respectively.

7. Through the corners of a square and perpendicular to its surface pass four wires, each carrying an upward current of 10 A. If the square is 2 cm on a side, what are the magnitude and direction of the net force per meter on one wire? This is the "pinch effect" observed in the parallel motion of charged particles in a plasma.

8. Two cylinders are wound with wire carrying a current. If the cylinders are placed end-to-end on the same axis and the current in each circulates the same way, will the cylinders attract or repel each other? Show that each cylinder behaves magnetically like a bar magnet.

9. An electromagnet has vertical parallel pole faces 5 cm in diameter and so close together that **B** is practically zero except between the faces.

    a)   A straight wire between the poles carries a current of 6 A from west to east and experiences an upward magnetic force of 0.36 N. What is the magnitude and direction of **B**?

    b)   What force would an electron experience if it moved at 10^7 m/sec from west to east between the poles?

10. A wire carries an electron current of 1 A. If an observer moves parallel with the wire at the same speed as, but in the opposite direction to, the average drift velocity of the electrons, what current will he observe in the wire? Explain.

11. A beam of alpha particles, which are known to be doubly charged helium ions, passes undeflected through crossed electric and magnetic fields. The electric field is produced by using parallel plates 2 cm apart with a pd of 200 V. The magnetic field is that inside a long helical coil with 318.4 ($=1000/\pi$) turns/m and carrying a current of 2.5 A. For such a helical coil or "solenoid" the magnetic induction B inside, as computed from Ampère's law, is given by the formula

$$B = \frac{4\pi k_m N I}{l},$$

where N/l represents the turns per meter of length and I the current. Compute E, B, and the speed of the alpha particles.

12. The alpha particles in Problem 11 are allowed to pass on into a region where their path is perpendicular to a second magnetic field of

0.5 N/A-m; the radius of their circular path is 40 cm.

    a)   Find q/m for the particles.

    b)   Assume that $q = 2e$ and find m.

    c)   Through what pd would one have to accelerate a helium ion from rest to give it the speed found in Problem 11?

AMPERE BALANCE **Experiment 13**
Ampère's Law and Calibration of an Ammeter

Object: To observe the magnetic force between adjacent currents and to calibrate an ammeter.

Theory: According to Ampère's law the magnetic force F_m between two parallel conductors each of length L, separated by a distance d, and each carrying the current I is

$$F_m = \frac{2k_m I^2 L}{d}. \tag{1}$$

The logical procedure followed has been to define k_m as 10^{-7} N/A^2 and to use this equation to define the *ampere*. By measuring F_m, L, and d, one may compute the value of I in a given case, pass this same current through an ammeter, and so calibrate the ammeter.

Procedure

Step 1. The Ampère balance* consists of a delicately balanced metal rod which may be located just above a similar fixed rod. Following a procedure similar to that of the Coulomb balance, one starts with a small mass m on the upper rod, which is adjusted so that with the weight mg acting downward, the two rods have a suitable separation, say about 1 cm. The weight is removed and the balance restored at the same separation by substituting for mg a magnetic force of attraction F_m just equal to mg. This is done by gradually increasing the current I through the two rods and watching the position of the movable rod through a telescope until it is seen to return to its original position. Then

$$F_m = \frac{2k_m I^2 L}{d} = mg. \tag{2}$$

---

* Purchasable from the W. M. Welch Mfg. Co., 7300 N. Linder Ave., Skokie, Ill. 60076.

Step 2. The length L and separation d must be measured. For the latter it is best to find the diameter of each conducting rod with calipers and the air gap separation of the rods with a telescope and scale. Then d is the center-to-center separation of the rods.

Step 3. By placing an ammeter in series with the Ampère balance, so that the same current I passes through each, one may check the scale reading of the ammeter for each value of I computed from Eq. (2). To obtain different currents one may alter either the mass m or the separation d. In so doing one may observe the relation between F_m and each of these factors.

Experiment 14   ELECTROLYSIS
Determination of the Electronic Charge e

Object: To measure the atomic unit of charge.

Theory: The smallest positive charge obtainable is believed to be that on an object that has lost one of its normal quota of electrons. We shall call this charge $+e$. If a neutral body gains an additional electron, the body will have a charge of $-e$. Millikan measured such charges on small oil drops, but this experiment is beyond the scope of this course. Chemists have proved that copper atoms in copper sulfate are bivalent, or that in a solution of this salt the sulfate ion takes two electrons from the copper atom and so each copper ion has a charge of $2e$. Here we shall measure the charge per copper ion and so find e.

Chemists have measured the atomic weight of copper to be 63.54 and 63.54 kg of copper is called 1 kg-mole. One kg-mole of any substance contains the same number (N_0) of atoms (what is N_0 called?); N_0 may be determined independently in a number of ways, such as by means of x-ray diffraction, radioactivity, sedimentation, etc., and the value found for it is 6×10^{26}. Knowing N_0, we can compute how many copper ions there are in a given mass (m) of copper.

By means of an electric current, one can cause m kg of copper to be electroplated out in a given time. From the definition relating current to charge, the charge q carried to an electrode by a current I flowing for t sec is

$$q = It.$$

By measuring m, one may compute how many copper ions it took to transport this charge and hence find the charge per ion. But each ion carried a charge $2e$, so e can be computed.

Procedure

Step 1. Derive the formula you will use to compute e in terms of N_0, m, I, t, and the atomic weight of copper.

Step 2. Clean the copper plate, rubbing it well with sandpaper and rinsing with water and alcohol. After the plate is *completely dry*, weigh it carefully on the balance. Place the plate in the battery jar nearly filled with copper sulfate solution.

Step 3. Connect the cell in series with the ammeter, a knife switch, a rheostat, and a source of current. Be sure the weighed copper plate is the negative electrode. Before closing the switch have your connections checked by the instructor. Now close the switch noting the time and regulating the current until it is approximately 1 A. Keep the current constant throughout the entire process by watching the ammeter carefully and changing the resistance if necessary. Allow the current to flow for 50 min. Remove the copper plate, after noting the final ammeter reading. Rinse the plate thoroughly and allow to dry before weighing. Weigh again very carefully to determine the amount of copper deposited.

Step 4. Compute e and compare your value with the accepted value of 1.6×10^{-19} C. Can you justify your percent error?

THE RATIO e/m FOR ELECTRONS   **Experiment 15**

Object: To measure the charge-to-mass ratio for electrons and compute the mass of an electron using the Bainbridge tube.*

Theory: We know from Ampère's law and the field concept that a moving charge is acted on by the magnetic field of neighboring moving charges. The latter will, in this experiment, be the electrons carrying a current I in two coils of wire. These coils will be so wound and situated that their separation equals the radius of each, and they are then called Helmholtz coils. For such a pair of coils the magnetic field in the center region between the coils is very uniform and, moreover, the strength of the field may be computed from the formula, based on Ampère's law,

$$B = \frac{32\pi nI}{\sqrt{125}a} \times 10^{-7}, \tag{1}$$

---

* This tube and the accompanying coils may be purchased from the W. M. Welch Mfg. Co. (see previous footnote). In it a slight amount of mercury vapor causes the electrons to leave a luminous track.

where B is the magnetic field in N/A-m (the mks-coulomb unit), n is the number of turns per coil, I is the current in amperes, and a is the mean radius of the coils.

When a charge e moves with speed v across a magnetic field **B**, a magnetic force $\mathbf{F}_m$ will act perpendicular to its motion. The particle then will travel in a circular path of radius R for which

$$e/m = v/BR. \tag{2}$$

The value of B may be computed from Eq. (1) and R may be measured, but how do we eliminate v? This is done as follows. We shall accelerate our electrons from rest to the speed v by pulling them through a known potential difference of V V. As 1 V $= 1$ J/C, the work done on the electron will be eV J. We equate this to the gain in kinetic energy and get

$$\tfrac{1}{2}mv^2 = eV. \tag{3}$$

Between (2) and (3) we can eliminate v and get

$$e/m = 2V/B^2R^2, \tag{4}$$

where V may be measured with a voltmeter.

Procedure

Step 1. Derive Eq. (4) and then substitute in it for B from Eq. (1).

Step 2. Inspect the apparatus and be sure of what each meter measures.

Step 3. Set the plate voltage at 30 V and close the filament switch. The plate current should be 5 to 8 mA, filament current 4.2 to 4.3 A; if this is not so *consult your instructor*, but *do not* try to change filament current yourself.

Step 4. Adjust the current in the coils so that the electron beam will be undeflected by the earth's magnetic field. This current reading must be subtracted from following readings.

Step 5. Increase the current in the coils until the electron beam is bent around so that its outer edge passes through the center of the prong nearest the filament; R is then 0.032 m.

Step 6. Repeat 5 for the next two prongs, for which $R = 0.039$ and 0.045 m, respectively.

204   *Step 7.* Set the plate voltage at 45 V and repeat Step 6.

Step 8. After compensating for the earth's field in the coil current readings, calculate V/I^2R^2 for each run and average your five values for this quantity. Use this average value of V/I^2R^2 to compute e/m.

Step 9. Compare your value with the accepted value of 1.76×10^{11} C/kg.

Step 10. From Experiments 14 and 15 compute your value for m, the mass of the electron.

9

9|1 INTRODUCTION   In Chapter 7 we considered electric charges at rest. The electric fields due to stationary charges do not change with the time and so have fixed patterns; they are called *electrostatic fields*. In Chapter 8 we investigated the magnetic fields due to steady currents in stationary conductors. Such magnetic fields are also unchanging in time. To complete our study of electric and magnetic fields we must also consider what happens when either type of field changes with time or is *unsteady*. In such circumstances, do any additional effects arise? We shall see that the answer is decidedly "Yes," and that these additional effects are the basis for our electric power and radio-television industries. The effects resulting

Faraday's and Maxwell's Laws of Electromagnetic Induction

from changing magnetic fields are summarized by Faraday's induction law and those resulting from changing electric fields are covered by a similar postulate of Maxwell. We shall discuss them in this order.

9|2 ELECTROMOTIVE FORCE

If an agent carries a positive charge around in an electrostatic field, the agent does work when moving the charge against the field and the field does the work when the charge moves with the field. Around a closed path work done against and work done by the field cancel when the field is an electrostatic one.

However, when an electric current flows around in a closed circuit (Fig. 9–1) a certain amount of heat is developed, and possibly light and mechanical energy as well are generated. We know that according to the conservation of energy principle, which has served us faithfully since its conception, energy must come from some source. Hence there must be a seat of nonelectrical energy in a closed current circuit. This may be (1) a battery, in which case the nonelectrical energy is chemical energy, (2) a generator, which requires mechanical energy to run it, (3) a photocell, in which light energy is used, (4) a thermocouple, which converts thermal energy into electrical energy, or (5) a microphone, which uses sound energy, and so on. In each case the nonelectrical energy is used to perform the work necessary to keep the current flowing around the circuit. In effect, the moving charges are driven around the circuit by a nonelectro-static electric field that does a net amount of work on each charge, even though the charges move around closed paths. The work done in this manner per unit charge is called the electromotive force (abbreviated to emf) around the closed path that a circuit may or may not occupy.

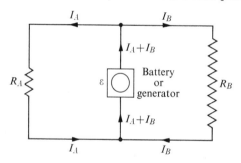

FIG. 9–1

Typical electric circuit. $\mathcal{E}$ is the source of emf (battery, generator, etc.). Picture R_A as representing the resistance of a light bulb and R_B that of a heating coil. $\mathcal{E}$ acts as a "pump" or "lift" for both currents, I_A and I_B. From Eq. (9–2), $\mathcal{E} = I_A R_A$ and $\mathcal{E} = I_B R_B$.

Since in the mks-coulomb system work per charge must be expressed in joules per coulomb, or volts, the volt is our unit for emf as well as for potential difference. The symbol for emf is the capital script letter $\mathcal{E}$.

If an electric current is to be maintained in a closed circuit, there must be an emf in the circuit, just as a closed water circuit needs a pump to maintain the flow against resistance and to maintain all the pressure drops in the various parts of the circuit. The emf of a circuit is usually localized, like the pump in a water circuit, but this need not be so. Another analogy is the typical ski area where one may see a continuous flow of skiers up the lifts and down the trails (Fig. 9–2). The lift corresponds to the source of emf in a current circuit. Each lift a skier rides gives him an increase in potential energy which he gradually loses as he coasts down the trail of his choice. The skier gains energy from the lift, and the lift gets its energy from the motor running it. The skier's energy is eventually dissipated as heat because on the trail his skis do work against friction, just as the conduction electrons in a wire do work against the electrical resistance of the wire.

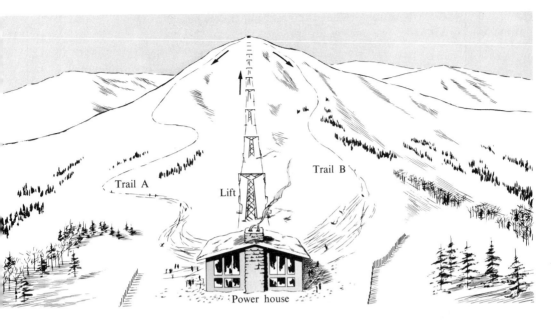

FIG. 9–2. In a ski area the lift supplies the skier with potential energy which, on the trail down, is dissipated through doing work against frictional resistance. (Compare with Fig. 9–1.)

Illustration. *Joule's law of electric heating*

Let a current I flow in a circuit containing the emf ε. Then the work done by ε when a charge q passes around the circuit is εq. The time rate of doing work, called the *power P*, is

$$P = \frac{\varepsilon q}{t} = \varepsilon \frac{q}{t} = \varepsilon I. \qquad \text{9–1}$$

Now suppose that the current I is proportional to ε (this is the case for the commonly used metallic conductors, but not for vacuum tubes or transistors), so that we may write

$$\varepsilon = IR, \qquad \text{9–2}$$

where R is a constant* for our special case. Then we may eliminate ε in Eq. (9–1) and write

$$P = I^2R. \qquad \text{9–3}$$

---

* When R is constant, or ε and I are proportional, the conductor is said to obey *Ohm's Law*.

With R constant we have the relation that

the power is proportional to the square of the current,

which is known as *Joule's law.* Of course one may always introduce and define R (it is called the *resistance*) by Eq. (9–2), but one cannot always assume that R is independent of the current. If R does depend on I, then P will not be proportional to I^2. The mks-coulomb unit for resistance is the volt per ampere, which is called the *ohm.*

9|3 FARADAY'S LAW. FUNDAMENTAL LAW X

The concept of electromotive force has been introduced because we shall see that Faraday's law may be most concisely stated by saying that a *changing magnetic flux through a circuit induces* (causes to arise) *an electromotive force around the circuit.*

From the work of Oersted, Ampère, and others, Faraday knew that electric currents could produce magnetic fields. With this knowledge he then began to look for the converse effect, i.e., currents produced by magnetic fields. While he found that a steady magnetic field passing through a stationary coil of wire connected to a galvanometer produced no current, he did observe a temporary current in the coil when the magnetic field linking it was changed. We now call this effect *electromagnetic induction,* and currents produced in this way are termed *induced currents.*

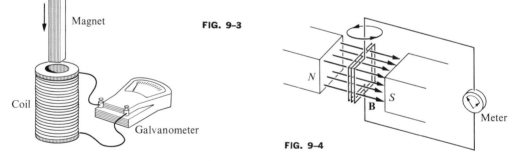

FIG. 9–3

FIG. 9–4

FIG. 9–3. A deflection may be produced in a galvanometer attached to a coil by thrusting a bar magnet into the coil.

FIG. 9–4. An emf is induced in a coil rotating in a fixed magnetic field.

Faraday proceeded to investigate all of the possible ways of producing induced currents and then he summarized his results in the law that bears his name. There are several different methods of inducing a current in a conducting circuit, such as (1) moving a bar magnet toward or away from the circuit (Fig. 9–3), (2) rotating the circuit in a fixed magnetic field, the principle of the electric generator (Fig. 9–4), (3) pulling the circuit out of a magnetic field, the principle used in a method of measuring magnetic fields, (4) moving another current circuit toward or away from the given circuit, and (5) changing the current in a second circuit placed adjacent to, or wound on the same iron core as, the first circuit, the principle of the electric transformer (Fig. 9–5). Other methods also exist.

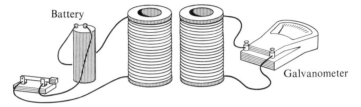

Battery

Galvanometer

FIG. 9–5. Momentary potential differences are established in the right-hand coil at the moment of opening or closing the switch in the left-hand circuit, or when, with its switch closed, the left-hand coil is moved toward or away from the right-hand coil.

As Faraday investigated all of these things the interesting question seemed to him to be: What do they all have in common? With the mind of a genius, Faraday found the answer.

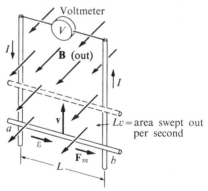

Voltmeter

$\mathbf{B}$ (out)

Lv = area swept out per second

FIG. 9–6. Motional electromotive force. Voltmeter deflects only when conductor ab is moving up and down.

A clue to this answer was obtained through the consideration of a special case. Suppose that a conductor of length L moves with the speed v at right angles to a magnetic field B, as in Fig. 9–6. The charges that are free to move in the conductor, ab, notably the free electrons in the case of a metallic conductor, are moving at right angles to a magnetic field and hence, according to Ampère's law, they must experience a magnetic force F_m given by

$$F_m = Bqv,$$

where q is the charge on the carrier. This force must be perpendicular to both the direction of the motion and the direction of the magnetic field, hence it must be *along the conductor*. This means that F_m will do work pushing the charges down the length L of the moving conductor. If the conductor is part of a closed circuit, the work per charge will constitute the emf ε contributed to the circuit by this moving branch. Let W be the work done on the charge q in the distance L. Then since work equals force times distance moved,

$$\begin{aligned} W &= F_m L \\ &= BqvL, \end{aligned}$$

and

$$\varepsilon = W/q = BvL. \qquad\qquad\qquad \textbf{9-4}$$

This is referred to as the expression for *motional emf*.

Example 1. A metal airplane with a wingspan of 60 m flies at 900 km/hr, cutting the vertical component of the earth's magnetic field, for which $B = 5 \times 10^{-5}$ N/A-m. What is the pd between the wingtips?

Solution.   Here $v = 9 \times 10^5$ m/3600 sec, and we have

$$\begin{aligned} \varepsilon &= vBl \\ &= \frac{9 \times 10^5 \text{ m}}{3600 \text{ sec}} \times \frac{5 \times 10^{-5} \text{ N}}{\text{A-m}} \times 60 \text{ m} \\ &= 0.75 \text{ N-m/A-sec} = 0.75 \text{ J/C} = 0.75 \text{ V}. \end{aligned}$$

One should realize that this value of ε would be much greater if the earth's field were comparable in strength with the magnetic fields of most magnets.

The clue to the general induction law that Faraday found in the above theory is the following. Since B may be considered to be the magnetic flux per unit area perpendicular to the direction of **B**, that is, the webers of flux per square meter (see Section 8–5), and since vL equals the rectangular area swept out across the field by the conductor in unit time, the product BvL is the flux in webers (Wb) cut by the moving conductor in unit time. If Φ is the flux cut in the time t, then Eq. (9–4) states that

$$\mathcal{E} = \frac{\Phi}{t}. \qquad\qquad \textbf{9–5}$$

Although Eq. (9–5) is equivalent to Eq. (9–4) as far as motional emf is concerned, Faraday realized that a generalization of Eq. (9–5) would account for all of the induction effects he had observed, including those listed above.

Faraday's law states that

the emf $\mathcal{E}$ induced in a circuit equals the time rate of change of magnetic flux through the circuit, or

$$\text{X.} \quad \mathcal{E} = \Delta\Phi/\Delta t, \qquad\qquad \textbf{9–6}$$

where $\Delta\Phi$ is the change in flux, in the time interval Δt, *due to any cause whatsoever*. This is our *Fundamental Law X*.

To appreciate the wide scope of this principle we must understand that the magnetic flux through an area A of surface perpendicular to the field **B** is defined as

$$\Phi = BA. \qquad\qquad \textbf{9–7}$$

Thus a change in either the area A of the circuit or in the concentration of flux B through a fixed circuit of constant area, or a combination of these changes, will result in a change in the total flux Φ through the circuit and give rise to a corresponding induced emf around the circuit.

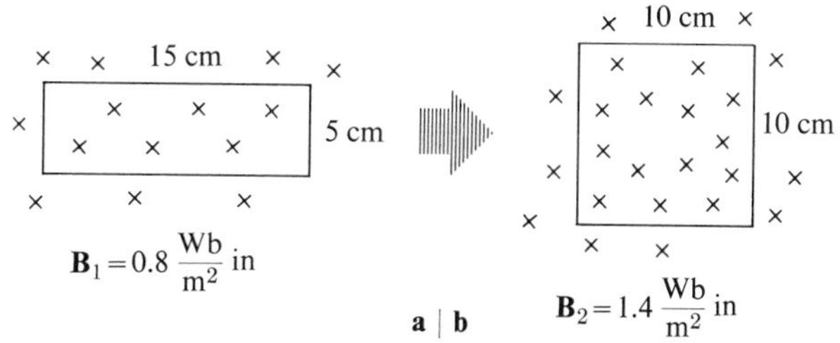

FIG. 9–7
Here the change in flux through the circuit and the resulting emf induced are due to both a change in the area and a change in the flux per unit area.

Example 2. A rectangular circuit whose dimensions are 5 cm by 15 cm lies perpendicular to a magnetic field **B** whose flux per unit area is $B = 0.8 \ \text{Wb/m}^2 = 0.8 \ \text{N/A-m}$ (Fig. 9–7). Suppose that in 0.5 sec the circuit is changed into a 10 cm by 10 cm square and B increased to 1.4 Wb/m^2, what would be the average induced emf during this 0.5 sec?

Solution. Initially the area is 75 cm^2, or $75 \times 10^{-4} \text{ m}^2$, so that the flux Φ_1 through the circuit is $75 \times 10^{-4} \times 0.8 = 6 \times 10^{-3}$ Wb. During the time interval $\Delta t = 0.5$ sec the area changes to $100 \text{ cm}^2 = 100 \times 10^{-4} \text{ m}^2$ and the flux to $\Phi_2 = 100 \times 10^{-4} \times 1.4 = 14 \times 10^{-3}$ Wb. Hence

$$\Delta \Phi = \Phi_2 - \Phi_1 = 14 \times 10^{-3} - 6 \times 10^{-3} = 8 \times 10^{-3} \text{ Wb}$$

and

$$\mathcal{E} = \frac{\Delta \Phi}{\Delta t} = \frac{8 \times 10^{-3} \text{ Wb}}{0.5 \text{ sec}} = 1.6 \times 10^{-2} \text{ V},$$

(since 1 Wb/sec is the same as 1 V).

Example 3. A coil of 100 turns and an area per turn of 0.02 m^2 lies on a table. In 0.5 sec the north end of a magnet is thrust toward the center of the coil, increasing the average B inside the coil from $5 \times 10^{-5} \text{ Wb/m}^2$ (about the value of the earth's field) to 0.4 Wb/m^2. Find the average induced emf for all the turns.

Solution. Here the flux per turn must be multiplied by 100 to obtain the flux linking all the turns. The initial flux Φ_1 is only one ten-thousandth of the final flux and so may be neglected. The final flux for all the turns is $\Phi_2 = 100 \times 0.02 \times 0.4 = 0.8$ Wb. Hence

$$\mathcal{E} = \Phi/t = 0.8 \text{ Wb}/0.5 \text{ sec} = 1.6 \text{ V}.$$

Example 4. *Alternating current generator.* Suppose that the coil in Fig. 9–4 is rotated round and round in a uniform magnetic field. The sense of the flux and induced emf will then reverse in direction every time the coil rotates 180°. The resulting current will also flow alternately one way and then the other. In fact both the emf and the current, if plotted against the time, will yield graphs that have the shape of a sinusoidal wave. Such alternating currents are called AC for short, in contrast with the DC (direct current) output of an electric storage battery, or dry cell. Note that it is relatively easy to generate AC, which is the type of current that our electric power stations generate.

9|4 LENZ'S LAW

This is really a part of Faraday's law. Expressions giving the magnitude of the induced emf $\mathcal{E}$ have been stated, but nothing has been said about the direction of $\mathcal{E}$.

214

In Fig. 9–6 the induced emf and the resulting induced current are counterclockwise when **B** is directed out from the page and the area of the circuit is decreasing. The flux through this circuit is decreasing in the outward direction. Now the induced current I produces its own magnetic field, and we may use the right-hand fist rule to compute the direction of this field. The result is that the magnetic field due to the induced current is also directed outward within the circuit. It is as though nature, through this induced field, tried to compensate for the reduction in the flux due to the applied field **B**. This turns out experimentally to be a general rule, so that we may say that

the direction of the induced emf is always such as to result in opposition to the change producing it.

This is *Lenz's law.*

As another example of the application of Lenz's law, consider a coil of wire to which a battery is suddenly connected (Fig. 9–8). Suppose that the battery starts a current flowing clockwise, as viewed by the observer. This current will give rise to a magnetic field whose lines will thread the coil and circle back outside it. Thus, as the current due to the battery builds up, there is a changing magnetic flux through the coil and this must result in an induced emf in the coil. What is the direction of this induced emf? Lenz's law tells us immediately that it must be counterclockwise, so as to oppose the building up of the current. Similarly, when the current in a circuit is broken, the induced emf seeks to keep the current from dying out, and this accounts for the sparking observed when switches are opened slowly. The induced emf in a circuit whose current is changing is called the *back emf,* since it always opposes the alteration in the current. It arises from the change in the current's own magnetic field, an effect referred to as *self-inductance.*

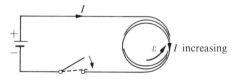

FIG. 9–8
Self-inductance. If the current *I* in the coil is increasing clockwise, the induced emf is counterclockwise, but if *I* is clockwise and decreasing, the induced emf is clockwise.

Were Lenz's law not true, an increase in current in a coil would result in an emf that aided the applied battery, thus increasing the current further, inducing more emf and further increases in the current, *ad infinitum.* This would be an unstable situation and one in which the conservation of energy principle would be disobeyed.

215

This sort of reasoning may be extended to other situations in which a system in equilibrium is displaced and the principle arrived at, which is really based on our Fundamental Laws IV and V, is the following.

When a system in equilibrium is disturbed, the equilibrium is displaced in the direction which tends to undo the effects of the disturbance.

This generalization of Lenz's law is called the *principle of Le Châtelier,* sometimes referred to as the "law of the cussedness of nature." It is, however, fortunate that it exists, or we would live in a very strange and unstable world indeed!

9|5 MUTUAL INDUCTANCE

Suppose that two circuits (call them 1 and 2) lie close together so that a current I_1 in circuit 1 causes magnetic flux to pass through circuit 2, as in Fig. 9–9. Changing the current in circuit 1 will alter the magnetic flux through both circuits and give rise to an induced back emf in circuit 1 and an induced emf and resulting current in circuit 2. The inducing of a current in circuit 2 in this manner was discovered independently by Faraday in England and by the American, Joseph Henry (1797–1879). The latter was, at the time, a schoolteacher at Albany Academy, although he was later to head the newly formed Smithsonian Institution in Washington, D.C. Faraday and Henry discovered what is now termed *mutual inductance.*

Illustration. *The transformer*

Suppose that two coils are wound on a common core or yoke of iron, as in Fig. 9–10, and that an alternating emf is applied to coil 1. The effect of the iron core is as follows.

The magnetic field due to the current I_1 magnetizes the iron, which means that there is an alignment of the spinning electrons in the iron. The effect is the same as though currents many times greater than I_1 circulated around in the iron in the same sense as I_1 circulates in its coil. The result is a much greater magnetic flux inside the core when it is made of iron than when it is made of, say, air, wood, or brass. Actually the contribution made to the flux by the spinning electrons in the iron may be over 100 times that due to the free electrons drifting through the external winding. A change in I_1 would then result in a correspondingly greater back emf. Thus the self-inductance of a coil with an iron core may be large.

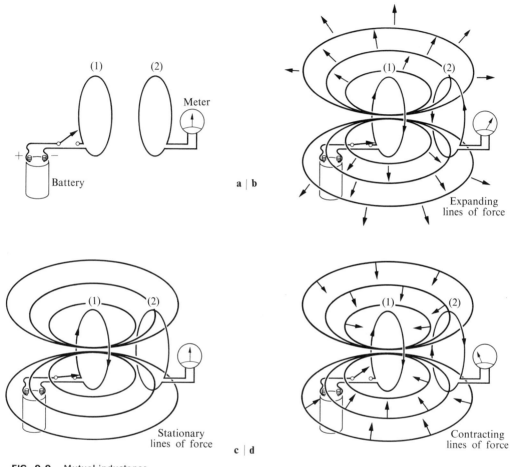

FIG. 9–9. Mutual inductance.

a| No current in either circuit.

b| Switch just closed; as current in circuit 1 increases clockwise, there is a temporary counter-clockwise current in circuit 2.

c| With steady current in circuit 1 there is no flux change or induced current in circuit 2.

d| Switch just opened; as current in circuit 1 dies out, there is a temporary induced current in the same sense in circuit 2.

In an iron-core ring or rectangular yoke the magnetic flux due to the iron, which is practically the whole flux, is constant around the closed core, so that the number of lines of flux through each circuit is proportional to the number of turns each circuit makes around the core. When changes in I_1 cause variations in the magnetization of the core, the changes in flux through the circuits, and hence their respective induced emf's, will

217

be proportional to the number of turns, N_1 and N_2, in each winding. Thus we have

$$\varepsilon_1/\varepsilon_2 = N_1/N_2. \qquad\qquad 9\text{–}8$$

The back emf ε_1 will oppose and approximately equal the amplitude of the alternating emf applied to circuit 1, while ε_2 will be the magnitude of the emf induced in circuit 2.

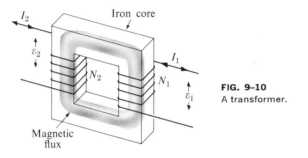

FIG. 9–10
A transformer.

Transformers are used to "step up" or "step down" the voltage in an alternating-current line. The effect on the current is the inverse of that on the voltage; since the power output of a transformer cannot exceed the power input, and since some of the input electrical energy is dissipated thermally, we may say that

$$\varepsilon_2 I_2 = \varepsilon_1 I_1 - \text{a dissipation factor.}$$

If the dissipation factor is negligible,

$$I_2/I_1 = \varepsilon_1/\varepsilon_2 = N_1/N_2. \qquad\qquad 9\text{–}9$$

Electric power is transmitted cross-country at high voltages and low currents so as to reduce the Joule heating effect, Eq. (9–3), and then the voltage is stepped down for the consumer of the electric power.

9|6 ENERGY STORED IN A MAGNETIC FIELD

When we build up a current in a circuit, an external source of emf must do work against the induced back emf. Work is also expended in heating the wire, but such work need not concern us here. The work in establishing a current in a circuit may be recovered by letting the current decrease back to zero, in which case the back emf will keep the current going a while after the external source of power has been disconnected. This suggests that associated with a current flowing in a circuit there is some form of stored energy. Where is this energy stored? We may think of it either as being associated with the charges comprising the current, or as being stored in the region containing the magnetic field resulting from the current. According to the first viewpoint, the stored energy is in the conductor and in a kinetic form, while in the field point of view we think of the energy as traveling out into space while the field is being established and returning to the wire when the field collapses.

What about electric fields? Do they also represent stored energy? Well, to produce electrostatic fields one must separate positive and negative charges, and this requires doing work against their Coulomb attractions. We may say that the charges possess potential energy due to their separated positions, but, after all, the charges are not altered in themselves and the creation of the field is the new thing. So it seems logical to think of an electric field as possessing stored energy which is recovered and converted into other forms when the field collapses.

Associating energy with electric and magnetic fields fits in with the theory of electromagnetic waves (Section 9–10) and with our experience regarding radio and television waves, which we *know* carry energy through empty space from transmitter to receiver.

9|7 FARADAY'S LAW APPLIED TO FIELDS IN FREE SPACE

So far we have considered the application of Faraday's law to closed circuits in which the induced emf produces an induced current. The presence of a conducting circuit is necessary as far as motional emf is concerned, which is not surprising since the expression for motional emf can be derived from Ampère's law for the force on a moving charge. When, however, one turns to where Faraday's law makes its new contribution, the postulate that a time-changing magnetic field may also induce an emf, then one finds that Faraday's law tells us something important about the interrelationships between electric and magnetic fields in general.

Consider a fixed circuit and suppose that the magnetic flux through it changes. Then an emf is induced in the circuit, which implies that a nonelectrostatic field is induced around the circuit. Now suppose that the conducting circuit is removed, but that the same change in flux occurs through the region. What happens? Without the presence of a conductor no induced current can arise, but what about the induced electric field?

It is fruitful to postulate that, whether conductors are present or not, *a changing magnetic field induces a nonelectrostatic electric field.* This is the field form of Faraday's law, our Fundamental Law X. Specifically it states that around a closed path through which the magnetic flux is changing at the rate $\Delta\Phi/\Delta t$, there will exist an induced electric field whose emf around the closed path is that given by Eq. (9–6), namely, $\varepsilon = \Delta\Phi/\Delta t$.

Illustration. *The betatron*

This is an interesting application of Faraday's law in atomic physics. We saw in Chapter 8 that an electron may be accelerated in an electric field, while a steady magnetic field may be used to deflect, but not to speed up, a beam of charged particles. We have just seen that a changing magnetic field produces an electric field; thus we have another method of speeding up charged particles. In the betatron this principle is used to accelerate electrons to very high energies.

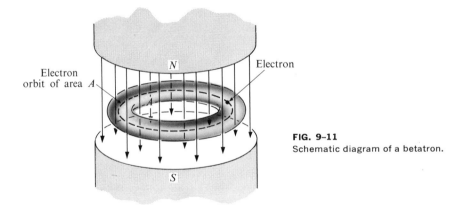

FIG. 9–11
Schematic diagram of a betatron.

Electrons from a hot filament are injected into an evacuated doughnut-shaped tube of mean radius R and area A (see Fig. 9–11). An alternating magnetic field is applied parallel to the axis of the tube. This field serves two purposes: (1) an emf tangential to the tube is produced by the changing magnetic flux, thus speeding up the electrons, and (2) a radial force (see Ampère's law) acts on the electrons as they move across the magnetic field, a force which may be used to keep the electrons in the circular path of the tube. To accomplish this, the magnetic induction must be a certain function of r, the distance from the axis, or the field must be produced by placing the tube between specially shaped poles of an electromagnet. Of course, the electrons can be kept in the tube and accelerated in one direction during only one-quarter of each cycle of the alternating field.

9|8 MAXWELL'S POSTULATE. FUNDAMENTAL LAW XI

Faraday was a brilliant *experimental* physicist, whereas the Scottish scientist James Clerk Maxwell (1831–1879) was a gifted *theoretical* physicist. Maxwell first read about and studied the work of Coulomb, Ampère, and Faraday and then proceeded to formulate their laws in terms of the vector calculus. He noted a certain lack of symmetry in the resulting equations, part of which he could attribute to the fact that nature provides us with free electric charges, but not with free magnetic poles. However, Maxwell was particularly struck by the fact that Faraday's law could be interpreted as saying that a changing magnetic field induces an electric field whereas the converse effect had not been observed or postulated. It would be simple if we could just say that Maxwell thereupon made the postulate that a changing electric field induces a magnetic field and added the corresponding term to his equations. In effect he did do this, but in the process he attempted to give his postulate a firmer foundation than the mere assumption that because one thing is true its converse must also be true. Furthermore, before Maxwell could express his postulate mathematically he had to determine whether an electric field changing at a given rate induces a strong or a weak magnetic field. Since no one had observed such an effect directly, before or during Maxwell's lifetime, it seemed pretty certain that in practical situations the magnetic field was weak. But how weak? Evidently Maxwell had to determine the value of a numerical factor. This factor turned out to be k_m/k_e, or $1/c^2$, where c is the speed of light.

How did Maxwell proceed? He showed that the equations expressing Coulomb's, Ampère's, and Faraday's laws were not compatible with the equation expressing the principle that electric charge can neither be created nor destroyed. He saw that the equations describing the behavior of electric and magnetic fields were consistent with the equation of charge conservation only for a steady state (in which all quantities are independent of the time) and not when the charge in a given region was changing with time. Maxwell found that to remove this inconsistency in the case of a nonsteady state it was only necessary to add to the equation summarizing Ampère's law an additional term, one involving the time rate of change of the electric field $\mathbf{E}$. With the addition of this new term, Maxwell's field equations became a concise summary of the laws of electromagnetism. The introduction of the new term amounted to postulating a law which, except for a numerical constant, is the converse of Faraday's law.

To express Maxwell's postulate numerically it is convenient to introduce the concept of *magnetomotive force* (mmf), to which we shall assign the symbol $\mathfrak{M}$. This is the magnetic counterpart of emf. Whereas emf is defined as the work to take a unit charge around a circuit or closed path, so mmf is defined as the work to take a unit pole around a closed path in a magnetic field. Unit pole may be defined as one on which a field $\mathbf{B} = 1 \text{ Wb/m}^2$ exerts a force of 1 N. (Remember that isolated magnetic poles have not been discovered, so that such a pole is a purely theoretical concept.) It is also useful to introduce the quantity *electric flux*, defined as the product of the electric field strength and the area of surface across which the field passes normally. Obviously electric flux is the electric analog of magnetic flux Φ. Let us represent electric flux with the symbol χ. We may now state the postulate derived by Maxwell as follows:

XI. $\mathfrak{M} = (k_m/k_e) \; \Delta\chi/\Delta t,$ **9–10**

that is,

around a closed path through which the electric flux is changing at the rate $\Delta\chi/\Delta t$, there will exist an induced magnetic field whose mmf around the path is $(k_m/k_e)(\Delta\chi/\Delta t)$.

This is our *Fundamental Law XI*.

9|9 COMPARISON BETWEEN MAXWELL'S AND FARADAY'S LAWS

The analogy between equations (9–6) and (9–10) is striking, yet there are important points in which the analogy breaks down. First, Eq. (9–10) contains the constant factor $k_m/k_e = 1/c^2$, which Eq. (9–6) does not have. Since $1/c^2$ is a very small quantity, one must conclude that, measured in mks-coulomb units, the magnetic fields induced by changing electric fields are numerically much smaller than the electric fields induced by changing magnetic fields. Furthermore, weak electric fields may be detected by the weak currents they can set up in a conducting circuit, whereas we cannot look for a weak "magnetic current" because neither free poles nor conductors for them exist. All of this explains why Maxwell's law was the last of the four fundamental laws of electromagnetism to be discovered.

A second important difference between Faraday's and Maxwell's laws concerns the directions of the induced fields. We saw in the last chapter that if $\mathbf{B}$ is increasing in strength toward the observer, the induced electric field is clockwise, as in Fig. 9–12(a). On the other hand, the magnetic field induced by a changing electric field is equivalent to that of a positive current in the direction of increasing **E**, so that one may apply the right-hand fist rule to determine the direction of the induced magnetic field; the result is that *in a region where* **E** *is increasing toward the observer, the induced magnetic field around the region is counterclockwise,* as in Fig. 9–12(b). We recall that when comparing the magnetic with the electric force between charges moving in parallel paths, we found the forces to be opposite in direction and that F_m/F_e also contained the factor $1/c^2$, so perhaps the conclusions above should not surprise us.

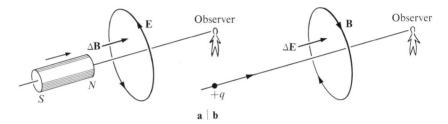

a | b

FIG. 9–12. Directions of induced electric and magnetic fields.

We may now summarize. The laws of Coulomb, Ampère, Faraday, and Maxwell tell us that, if we regard the fields of magnets as magnetic fields due to charges (electrons) in motion (spinning), then there are two basic ways in which electric fields may be obtained and two basic ways of producing magnetic fields, as shown in Table 9–1.

TABLE 9–1

Electric fields are associated with	Magnetic fields are associated with
Charges at rest	Charges in motion
Changing magnetic fields	Changing electric fields

9|10 ELECTROMAGNETIC WAVES

We turn now to the great prediction in Maxwell's theory. Coulomb, Ampère, and Faraday arrived at their fundamental postulates as a result of experimentation and the inductive process of seeking a statement that summarizes a host of different observed facts. Maxwell, on the other hand, did not summarize any new experimental facts, but he did make predictions that were eventually found to be true and thus justified his theory. Let us consider in a qualitative way how his theory predicted that electric and magnetic fields may be propagated through space as transverse waves.

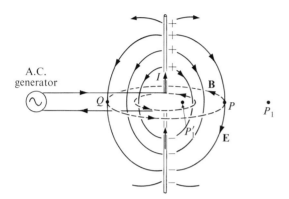

A.C. generator

FIG. 9–13
Antenna approaching state of maximum charge.

A simple generator of electromagnetic waves is shown in Fig. 9–13; a source of alternating emf is connected to two metal rods that form an antenna. Suppose that at the instant shown, electrons are flowing downward, or the direction of the positive current **I** is upward, and that positive charges are accumulating on the upper wire and negative charges on the lower wire. The electric field **E** will be similar to that of a dipole (Fig. 7–9) as shown by solid lines. The magnetic field **B** will be in circles about the antenna, and its direction, according to the right-hand fist rule of Ampère, will be into the page at P and out at Q, as shown by the dashed curve. Thus at P and Q the fields **E** and **B** are at right angles.

Next consider how these fields are changing with the time. The charges on the wires are increasing so that, according to Coulomb's law, their electric field will spread out farther, say to P_1. Since the charges on the wires are approaching their peak values, after which they will decrease in magnitude, the current I must be decreasing toward zero, after which it will be downward. A decreasing I means that at P the magnetic field **B** is decreasing with the time; according to Faraday's and Lenz's laws, if **B** is in and decreasing (or out and increasing), there must be an induced electric field **E'** directed in loops about the changing magnetic field. The direction of this new electric field will be downward at P_1 and upward at P_1', so that it will reinforce **E** at P_1 and oppose it at P_1'. Thus the region where the electric field is strongly downward moves out away from the wire, as in Fig. 9–14.

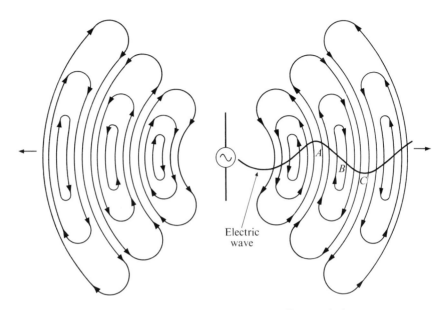

FIG. 9-14. Pulses of electric field spreading out from an antenna with oscillating polarity. At *A* the magnetic field is out; at *B* it is zero; at *C* it is in.

Finally, a similar consideration of the magnetic effect of the changing electric field shows that an inwardly directed wave of the magnetic field accompanies the downward wave of the electric field that is in motion away from the antenna. This wave of inward **B** and downward **E** will be followed by one of outward **B** and upward **E**. As the polarity of the antenna is periodically reversed, a continuous wave of electric and magnetic fields will, according to the four laws of electromagnetism, be radiated outward. These waves are transverse with respect to both **E** and **B**. The direction of propagation is the direction of advance of a right-hand screw rotating from **E** to **B**, as shown in Fig. 9–15.

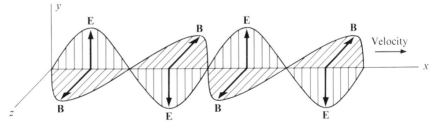

FIG. 9-15. Relationship between the directions of the electric field, the magnetic field and the direction of propagation in an electromagnetic wave.

Out in space, far from the transmitting antenna, electromagnetic waves are self-propagating. The changing electric field induces a new magnetic field which, in turn, also changes and induces a new electric field. The ingredients necessary for the propagation of a mechanical type of wave are inertia in the transmitting medium and a restoring force when the medium is disturbed. Electromagnetic waves do not require a material medium for their propagation because they contain, in a sense, the same necessary ingredients as built-in properties, according to Faraday's and Maxwell's laws. Mechanical waves consist of oscillating particles whose energies are both kinetic and potential. By analogy, one may think of the electric energy of a charged conductor as potential energy stored in its electric field, and the energy stored in the magnetic field due to a current in a coil as analogous to kinetic energy. Thus an electromagnetic wave contains the equivalent of KE and PE.

When electromagnetic waves pass into matter they encounter electrons, such as the bound electrons orbiting in atoms and the free electrons in conductors. As the alternating electric field $\mathbf{E}$ passes an electron, the electron is urged alternately one way and the opposite way. When $\mathbf{E}$ is upward the electron is urged downward, and when $\mathbf{E}$ is downward the force on the electron is upward. The moving electron is also affected by the magnetic field $\mathbf{B}$ in the wave. The result of the interaction between an electromagnetic field and the electrons in matter is that the electrons absorb energy from the field. This absorbed energy may (1) be turned into internal thermal energy of the absorber, (2) be reradiated in different directions, or (3) it may give rise to alternating currents in a receiving antenna and connecting circuit. For process (3) to be most efficient, the receiving circuit must be "tuned" to resonance with the frequency of the waves that are being sent out and picked up, i.e., the frequency of the transmitter.

Theoretical considerations led Maxwell to predict that electromagnetic waves such as those just described should be propagated through space as transverse waves with the speed

$$v = \sqrt{k_e/k_m},$$

that is, with the speed of light, and that these waves should have the other properties of light except, perhaps, visibility. Maxwell was thus led to make the very educated guess that electromagnetic waves and light are one and the same thing, or, to put it another way, that electromagnetic waves furnish a good model for describing the behavior of light while it is being propagated from where it is emitted to where it is absorbed.

Experimental confirmation of Maxwell's postulate did not come until 15 years after he published his work. In 1887–1888 another brilliant experimenter, Heinrich Hertz of Germany, succeeded in producing and detecting electromagnetic waves of the kind we now call radio waves. By 1901 Marconi had sent radio waves from the outer cliffs of Cape Cod across the Atlantic Ocean. Now, of course, we use electromagnetic waves not only for radio transmission, but in television, radar, space communication, and the study of molecular vibrations.

PROBLEMS

1. Suppose that in a conductor $I = 1$ A when $\varepsilon = 5$ V and $I = 2$ A when $\varepsilon = 10$ V. Find the electric energy expended in 1 min when

    a) $I = 1$ A, and   b) $I = 2$ A.

    c)   Is Joule's law obeyed here? Explain.

2. Repeat problem 2 for another conductor for which $I = 1$ A when $\varepsilon = 5$ V and $I = 2$ A when $\varepsilon = 12$ V.

3. Compute the resistance R when $I = 1$ A and when $I = 2$ A for each of the conductors in the preceding two problems. Which conductor definitely does not obey Ohm's law? Joule's law?

4. Suppose that in Fig. 9–1 $\varepsilon = 110$ V, $R_A = 55$ ohms, and $R_B = 22$ ohms. Find

    a) I_A, I_B, and the current from the generator, and

    b)   the power developed in R_A and R_B respectively. Is more power spent in the lower or in the higher resistance?

5. In Fig. 9–6 let ab be 0.5 m, $B = 0.2$ Wb/m^2 $= 0.2$ N/A-m, $v = 40$ m/sec, and $I = 3$ A. Compute

    a)   the motional emf,

    b)   the power generated,

    c)   the force on ab, and

    d)   the work done per second against this force.

6. A coil has 100 turns, each of 0.5-m^2 area. It is placed horizontally in a region where the earth's magnetic field has a vertical component of 5×10^{-5} Wb/m^2. Compute the induced emf if the coil is turned over in

    a)   0.1 sec, and   b)   0.2 sec.

227

7. In problem 6 assume that $R = 20$ ohms and compute the induced current and induced charge in each case.

8. A straight wire of length l is rotating about one end at n revolutions/sec in a plane perpendicular to a magnetic field **B**. Find, in terms of l, n, and B, the emf induced between the ends of the wire.

9. Use Lenz's law and the right-hand fist rule to

a) find whether the induced emf in the example illustrated in Fig. 9–7 is clockwise or counterclockwise,

b) verify the direction of the induced current in circuit 2 of Fig. 9–9(b) and (d).

10. A transformer has 5 turns in its primary winding and 100 turns in its secondary. If the primary voltage is 110 V and the primary current is 10 A, find the secondary voltage and current, assuming

a) no dissipation of electric energy, and

b) a 10% loss of electric energy in the transformer.

11. An electromagnetic wave strikes a surface at right angles to the direction in which the wave is traveling. If the wave carries energy to the surface at the rate of 1.4×10^3 W/m^2, which is about the intensity of sunlight above our atmosphere, how much energy must each cubic meter of the electromagnetic field contain?

12. Describe the patterns of the electric and the magnetic fields near a bar magnet that is insulated and charged positively. If, as Maxwell's theory predicts, energy is propagated in the direction of advance of a right-hand screw rotating from **E** to **B**, will such a charged magnet gain or lose energy? Explain.

13. A steady current flows up in a long straight wire in which **E** is in the upward direction. Does the wire gain energy from, or lose energy to, the electromagnetic field?

Experiment 16   FARADAY'S LAW
Lenz's Rule

Object: To verify Faraday's induction law and Lenz's rule for the direction of the induced emf.

Theory: Faraday discovered that a changing magnetic field generates an electric field, which may be used to drive a current. If the magnetic flux Φ

through a coil of wire changes by $\Delta\Phi$ in the time Δt, then the emf induced in the coil is given by $\Delta\Phi/\Delta t$. In a circuit of resistance R the induced current I will be

$$I = -\frac{1}{R}\frac{\Delta\Phi}{\Delta t}. \tag{1}$$

The minus sign here is a reminder that I itself is in such a direction that its own magnetic field will oppose the change $\Delta\Phi$ producing I; this is *Lenz's rule*.

The flux change $\Delta\Phi$ may be produced in many ways, several of which you will try. To verify Eq. (1), you will vary both $\Delta\Phi$ and Δt. To verify Lenz's rule, you will have to establish clearly the directions of $\Delta\Phi$ and of I. Knowing the direction of I, you may use the right-hand fist rule to find the direction of I's magnetic field, which should come out opposite to the direction in which Φ is increasing, or the same as the direction in which Φ is decreasing.

Procedure

Step 1. Note how the coils are wound. Is there an extra half turn over the top, so that the entering current goes first over the top, or does the entering current have to go first around the bottom?

Step 2. Connect a high resistance in series with a dry cell and table galvanometer. Put a tap switch in the circuit. Determine the polarity of the cell. Close the switch momentarily and note the direction of the throw of the arm of the galvanometer when the current enters at a given terminal. Now you can tell from the right or left deflections of the galvanometer the direction of the induced currents in the remainder of the experiment. (Disassemble circuit.)

Step 3. Connect one coil to the galvanometer with a resistance box in series. The resistance box is used to control the sensitivity of the galvanometer. Use the resistance box only to keep the deflections from going off scale and record resistance used in such cases.

a) Bring up the N-pole of the magnet slowly and take it away slowly.

b) Repeat (a) with the S-pole.

c) Repeat (a) with the N-pole but bring it up faster and take it away faster.

Note direction of current in the coil in each case and compare deflections (or currents) in (a) and (c).

Step 4. With one coil connected to the galvanometer and resistance box and the other coil to the dry cell

a) bring one coil up to the other fast and take it away fast;

b) reverse the direction of the current and repeat (a).

Note the direction of current flowing through the coils.

Step 5. Connect as in Step 4 but put a switch in the dry cell circuit and place the coils together. Make and break the current flowing in the dry cell circuit.

Step 6. Repeat Step 5 with an iron core through the coils.

Step 7. Write a brief summary of the results, indicating how Faraday's law and Lenz's rule were verified. Include diagrams showing clearly the directions of $\Delta\Phi$, I, and the magnetic field due to I.

Experiment 17 MICROWAVES
Optical Properties of Electromagnetic Waves

Object: To observe that electromagnetic waves produced electrically may be made to undergo reflection, interference, and diffraction, and to measure the wavelength of these waves.

Theory: With a special vacuum tube known as a klystron it is possible to generate electromagnetic waves in the centimeter region, i.e., waves with a wavelength λ of a few centimeters. Like visible light, these so-called *microwaves* travel with the speed of light and may undergo reflection, diffraction, and interference. However, a wavelength of, say, 3 cm is much greater than that of visible light (for which λ ranges from 4×10^{-5} cm to 8×10^{-5} cm). This means that diffraction and interference experiments may be performed more easily and on a more familiar scale if microwaves are used in place of visible light.

Your microwaves will be emitted from a small horn (the source S) and picked up by a similar receiving horn (R). In the latter, the signal is converted into an alternating electric current which may be fed into earphones or into a more sensitive detector such as an oscilloscope. A plane sheet of metal serves as a very good reflecting mirror for microwaves, and a coarse wire grid behaves like a half-silvered mirror, allowing part of the beam to pass through and reflecting the rest. Two metal sheets with straight edges may be placed close together to form a suitable slit.

Procedure: You may, with microwaves, repeat these three famous optical experiments, namely (1) thin film interference, (2) the Michelson interferometer, and (3) Young's double slit interference experiment.

Step 1. Place two coarse wire grids between the source S and receiver R and perpendicular to SR, as in Fig. 9–16. It is preferable to mount everything on a meter stick or optical bench. Move one grid slowly and note positions for which the signal received is a maximum. Average the distances between these maximum positions. This should be half the wavelength of the microwaves used.

The explanation is that the second "half mirror" or grid reflects waves back to the first grid which then reflects part of these toward R. If these latter waves join the unreflected waves in phase, a maximum signal is received. Now if the separation of the grids is increased by $\lambda/2$, the doubly reflected waves travel one wavelength farther than before and so still come out in step with the unreflected waves, giving another maximum. The same condition holds for visible light reflected from the top and bottom surfaces of thin films, such as in oil slicks and soap bubbles.

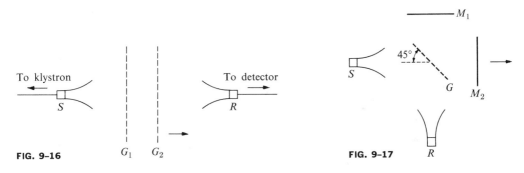

FIG. 9–16. **FIG. 9–17.**

FIG. 9–16. Thin film interference experiment for microwaves.

FIG. 9–17. The Michelson interferometer experiment with microwaves.

Step 2. Set the two reflecting plates M_1 and M_2 and the grid G as in Fig. 9–17. This constitutes a Michelson interferometer, which is described in detail early in the next chapter. Move either M_1 or M_2 slowly and note positions for which maximum reception is picked up at R. Again the average distance moved between successive maximum positions should be $\lambda/2$, because if M moves back that much then the wave reflected from it must travel that distance and back, or λ cm farther than before.

Step 3. If possible, see whether you can duplicate Young's two-slit experiment (see Fig. 6–21). Due to the low intensity of the diffracted waves a sensitive detecting device is required.

Step 4. Compute the frequency of your microwaves from the measured wavelength and the knowledge that they travel with the speed of light.

10|1 INTRODUCTION The fundamental laws of physics that we have
considered so far all date from before the year
1900, whereas those discussed in this and the
following chapters were only discovered in the
present century. The physics of the period
prior to 1900 is frequently referred to as
"classical physics," and we have seen that it
succeeded in discovering the laws and building
the theories in terms of which the ordinary
occurrences of our physical world can be explained. The reason that so much could be
accomplished without the use of the relativity
and quantum principles of 20th-century
"modern physics" is that these latter principles are only significant when we extend our

The Relativity Principle

investigations from the realm of man's ordinary macroscopic world to that of the microscopic world of atoms and nuclei, or to the astronomical world of the cosmos.

It is not that the relativity and quantum principles fail to apply to such everyday phenomena as the motion of a baseball or a car, for they do apply, but the point is that they do not *have* to be applied. We shall see that when the expressions of the relativity theory differ from those of Newtonian theory, the former reduce to the latter when the speed of the material objects with which we are concerned is considerably less than c, the speed of light. The newer theory embraces the older one as a special case; this case, however, includes most common occurrences because the speed of light happens to be very *great*. If the speed of light were infinite, those effects peculiar to the relativity theory would not occur; we would then always be able to use Newtonian theory, which is simpler and easier to apply.

The quantum theory bears the same relationship to classical theories; its fundamental constant is Planck's constant h and it is because h is so *small* that quantum effects do not appear in the macroscopic world. If h were zero, we would not need the quantum theory at all.

10|2 INERTIAL SYSTEMS

By a *reference system* we shall mean a frame of reference, such as a set of x-, y-, z-axes and a clock, stationary with respect to the axes, for measuring the time t at which an event occurs. By an *inertial system* we shall mean a reference system S relative to which Newton's laws of motion take their customary form, or any other system S' that is moving relative to S with *unaccelerated* motion. Since the earth is rotating, it does not constitute a true inertial system, but in laboratory experiments the effects of the earth's rotation are negligible and the ground or walls of a building serve sufficiently well for an inertial system. Newton believed an ideal inertial system to be one that is absolutely at rest. A large group of so-called "fixed stars" are relatively at rest, and these are sometimes taken as defining an "absolute" system. However, we shall see that the term "absolutely at rest" has no meaning and that all motion is relative. It is unimportant to ask which one of two or more inertial systems is at rest.

Suppose that a certain law is deduced and its predictions are found to hold true experimentally by observers in one inertial system, say that of the laboratory. Now consider a second set of observers who are investigating the same law on a spaceship that is moving at a large but constant velocity relative to the laboratory. If both sets of observers conclude that the law, in the same form, holds true, then the law is said to be *relativistically invariant*. Newton believed that his principles possessed this property.

10|3 THE PRINCIPLE OF RELATIVITY. FUNDAMENTAL LAW XII

We are familiar with the situation in which two trains pass each other slowly and steadily on parallel tracks and a passenger looks out of the window of one train and sees only the other one; the passenger cannot say what the speed of either train is relative to the ground, he can only measure the relative motion of the two trains. Or suppose, for example, that we are enclosed in a large box without windows, such as a freight car or an elevator, and that this box can move smoothly and quietly relative to the ground; again we will not be able to detect or measure the motion if it is *unaccelerated*. Of course accelerated motion is detectable because when a car speeds up one is thrown backwards, and when a car is centripetally accelerated on a curved track one is thrown to the side; in an elevator we are aware of acceleration because of the apparent change in our weight.

Einstein's *special theory of relativity*, which is what we shall discuss, was proposed in 1905 and deals with unaccelerated motion. Later he developed the more *general theory* that is concerned with accelerated systems and gravity, a theory that requires for its understanding an advanced training in mathematics as well as physics.

In a generalization of what has been said above regarding our inability to detect unaccelerated motion, Einstein made the following postulate:

There is no physical experiment that can be performed, relative to one inertial system only, by means of which the uniform velocity of that system relative to another inertial system can be detected.

This is Einstein's *principle of relativity* and our *Fundamental Law XII*. We shall break this principle down into two parts.

a | **The speed of light** *in vacuo* **as measured relative to any inertial system is always the same.**

It is independent of the motion of the source and of that of the observer. The speed, relative to the ground, of a bullet fired from a moving plane depends on the speed of the plane; the speed of sound, relative to the ground, depends on the speed of the air that carries the sound; but nothing "carries light" in empty space. We shall show that this postulate is verified experimentally by the Michelson-Morley experiment and that it leads mathematically to the Lorentz space-time transformation equations.

b | **The fundamental laws of physics have the same form, or are invariant, for all inertial systems.**

Some time before Einstein announced his theory of relativity, H. A. Lorentz showed that the equations expressing the laws of electromagnetic fields are invariant if one assumes the same space-time transformations referred to in Section (a). Einstein proved that the invariance of Newton's principles also follows from the same space-time transformations if one assumes that the mass of a body varies with its energy. The consequences of this assumption have been fully tested experimentally—witness the atomic bomb.

The Lorentz space-time transformation equations plus Einstein's mass-energy equivalence relation, $E = mc^2$, thus summarize Einstein's principle mathematically.

10|4 THE MICHELSON-MORLEY EXPERIMENT

To make the meaning of this experiment clearer, the following analogy is frequently presented.

Suppose that in Fig. 10–1, A and B are two points on the opposite banks of a stream which is flowing with the speed v relative to the ground. Let l be the width of the stream and let C be a point a distance l upstream from A. We imagine a race between two identical swimmers, each of whom swims with the speed c in still water, one swimmer being told to swim across the stream to B and back to A, while the other must swim upstream to C and then back to A. How will the times of the two swimmers compare?

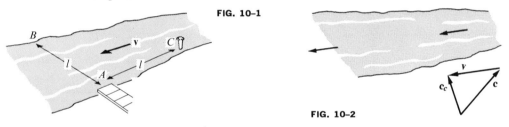

FIG. 10–1

FIG. 10–2

FIG. 10–1. Race between two swimmers in a flowing stream.

FIG. 10–2. The swimmer's effective velocity c_c across stream is less in magnitude than his speed c in still water.

Consider the first swimmer, who must cross the stream. In order to allow for the drift current v, he must aim diagonally upstream so that the component of his effective velocity across the stream, say c_c, is only one component of the vector c, the other being equal and opposite to the current v (see Fig. 10–2). Then since $c^2 = c_c^2 + v^2$ (see Chapter 3 for vector addition), we have

$$c_c = \sqrt{c^2 - v^2}.$$

The distance across and back is $2l$, hence the time t_1 for this swimmer will be $2l/c_c$ (distance divided by effective crossing speed), or

$$t_1 = \frac{2l}{\sqrt{c^2 - v^2}}.$$

10–1

The other swimmer makes slow progress upstream, his effective speed being $c - v$, but he has the current with him on the return leg, when his resultant speed is $c + v$. His total time t_2 will thus be $l/(c - v)$ up plus $l/(c + v)$ back, or

$$t_2 = \frac{l}{c - v} + \frac{l}{c + v} = \frac{2lc}{c^2 - v^2}. \qquad \textbf{10-2}$$

To compare t_2 with t_1 let us divide the latter by the former. Since the $2l$ factors cancel out, we have

$$\frac{t_1}{t_2} = \frac{c^2 - v^2}{c\sqrt{c^2 - v^2}} = \frac{\sqrt{c^2 - v^2}}{c} = \sqrt{1 - \frac{v^2}{c^2}}. \qquad \textbf{10-3}$$

We see that t_1/t_2 *is less than unity*, so that t_1 is less than t_2 and the first swimmer wins.

Let us assume that the two swimmers are accurately timed. Suppose (this is a purely hypothetical case) that the two times were found to be exactly the same, what explanations might be offered? We might conclude that (1) the ratio v/c is zero, within the limits of measurement, or (2) that the distance from A to C was inaccurately measured and that it is actually less than the distance across the stream. Conclusion (1) implies either no appreciable current ($v = 0$), or very fast swimmers ($c \rightarrow \infty$). Conclusion (2) would necessitate a shortening of the distance $\overline{AC}$ to $l\sqrt{1 - v^2/c^2}$, where l is the width of the stream; with $l\sqrt{1 - v^2/c^2}$ in place of l in the expression for t_2, we would have

$$t_2 = \frac{2lc\sqrt{1 - v^2/c^2}}{c^2 - v^2} = \frac{2l\sqrt{c^2 - v^2}}{c^2 - v^2} = \frac{2l}{\sqrt{c^2 - v^2}} = t_1.$$

The reason that t_1 and t_2 are different is that the current, which carries the swimmers along, affects them differently. In the same way, sound waves carried by the air travel faster with the wind than across or against the wind. In the time of Maxwell much thought was given to the question of whether there was any medium which carried light. Some physicists suggested that there was such a medium, pervading all space, which they called the *ether*, and they regarded it as an absolute frame of reference. If such an ether existed, the earth could not always be at rest relative to the ether, since in six months' time the motion of the earth in its orbit is reversed. Suppose that the earth is moving through the ether; then this relative motion, called *ether drift*, should affect the motion of light just as the current of the stream was seen to alter the effective speeds of the two swimmers.

Maxwell suggested a crucial experiment for testing the ether theory. This experiment was performed by Michelson, who was shortly aided by Morley, and it is called the *Michelson-Morley experiment.* The apparatus employed was essentially a Michelson interferometer floating on a pool of mercury. The interferometer, diagrammed in Fig. 10–3, is the optical analog of the case of the stream and the two swimmers. A beam of light from a source Q strikes a half-silvered glass plate G, so that part of the light is reflected to the mirror M_1 and part passes through G and is reflected by the mirror M_2. Upon returning to G the two beams are again partly reflected and partly transmitted. The transmitted part of the beam from M_1 joins the reflected part of the beam from M_2 and the combined light passes into the telescope T. The optical lengths l of the two paths are made the same. Then if there is an ether drift v parallel to GM_2, the light from M_2 will be, with respect to phase, a little behind that from M_1. A pattern of interference fringes is observed through the telescope T.

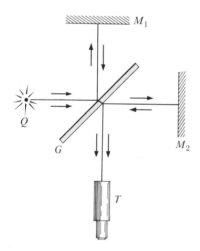

FIG. 10–3
The Michelson-Morley experiment.

The crucial part of the experiment comes when the whole apparatus is rotated 90° about a vertical axis, so that any ether drift originally parallel to GM_2 will now be parallel to GM_1, causing a shift in the relative phases of the two combining beams. During the rotation the observer must watch for a resulting shift in the interference pattern seen through the telescope. Taking for v the earth's orbital velocity of 30 km/sec, Michelson and Morley computed for their apparatus a shift of about half a fringe, according to the ether theory. Actually they found no significant shift, although they believed that a shift of $\frac{1}{100}$ of a fringe could have been observed. This experiment has been repeated many times since Michelson first attempted it in 1880, and it is now generally agreed that the earth's motion through space cannot be detected as a motion through the ether.

As in the case of the two swimmers, the null result in the Michelson-Morley experiment may be interpreted as indicating (1) that the drift current is effectively zero, or (2) that distances parallel to the drift current are effectively shorter by a factor $\sqrt{1 - v^2/c^2}$ than they appear to be to an observer moving with the apparatus. As we have said, the earth moves in different directions at different times of the year, and therefore if there were an ether throughout space the earth could not always be at rest relative to it. The possibility that the earth carries with it an envelope of ether, stationary relative to the earth, has been ruled out by other experimental evidence. Conclusion (1) then necessitates giving up the ether drift hypothesis. The Irish physicist Fitzgerald suggested possibility (2), which amounts to the following.

Suppose that an observer measures the length of an object to be l' when he is moving with the object, but he finds the same length to be $l = l'\sqrt{1 - v^2/c^2}$ (which is less than l') when the object is moving past him longitudinally with a speed v. In the Michelson-Morley experiment the observer is moving with the apparatus and he claims that the apparent length of the longitudinal path is equal to the length of the transverse path. An observer at rest relative to the fixed stars would say that the longitudinal path is shorter by the factor $\sqrt{1 - v^2/c^2}$, so that the times for the two light paths would be the same. Since any measuring stick placed alongside the Michelson apparatus will undergo the same contraction, an observer moving with the apparatus cannot detect the motion of the apparatus relative to the fixed stars.

While the Fitzgerald contraction theory does explain the null result of the Michelson-Morley experiment, this theory certainly seems rather artificial and lacking in scope. Einstein showed that by using his postulate that the speed of light is independent of the motion of the observer he could not only immediately explain the result of the Michelson-Morley experiment but make many interesting predictions, all of which have been verified experimentally. Einstein's theory eliminates the need for postulating an ether or the concept of absolute motion relative to the fixed stars. Fitzgerald's proposed contraction becomes a corollary of the relativity theory.

We shall next employ Einstein's postulate to derive the Lorentz space-time transformation equations and then discuss some of the many interesting corollaries that follow as a consequence.

10|5 RELATIVITY DERIVATION OF THE LORENTZ SPACE-TIME TRANSFORMATION

Consider two inertial systems, S and S'. Take the x-axis of S to coincide with the x'-axis of S' and let S' be moving relative to S with the speed v in the x-direction (Fig. 10–4).

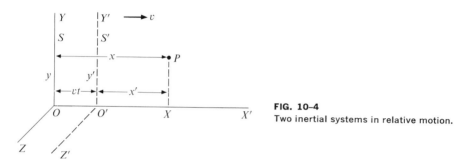

FIG. 10–4
Two inertial systems in relative motion.

Let the two sets of axes coincide at the time we shall take to be $t = 0$ and $t' = 0$; starting at this time, let a spherical light wave spread out from the origin. At the time t in S this wave will reach some point P whose coordinates are x, y, z. Let the wave reach P at the time t' as measured by observers in the system S' and call the coordinates of P in this system x', y', z'. Since speed of propagation equals distance traveled divided by the time taken, we have

$$c = \frac{\sqrt{x^2 + y^2 + z^2}}{t},$$

or

$$x^2 + y^2 + z^2 - c^2 t^2 = 0. \qquad \text{10–4}$$

Similarly, we may write for observations made in S',

$$x'^2 + y'^2 + z'^2 - c^2 t'^2 = 0, \qquad \text{10–5}$$

where the c is *not* primed since, according to Einstein's postulate, c is the *same* in S' as in S.

From symmetry considerations it seems reasonable to try putting $y = y'$ and $z = z'$, since the relative motion of the systems is in the x- or x'-direction and not in the y- or z-directions. Combining Eqs. (10–4) and (10–5), we then get

$$x^2 - c^2 t^2 = x'^2 - c^2 t'^2. \qquad \text{10–6}$$

We seek the transformation relations between x', t' and x, t which satisfy this equation by reducing it to an identity.

In accordance with Galilean-Newtonian theory, we would naturally assume the transformation equations to be (see Fig. 10–4)

$$x' = x - vt, \qquad t' = t, \tag{10-7}$$

which are called the *Galilean*, or *Newtonian*, *space-time transformation equations*. These equations are in agreement with observations involving familiar moving systems, such as trains or airplanes, for which $v \ll c$.* Does not the primed system move a distance vt to the right in the time t? One can see by substitution that these equations do not satisfy Eq. (10–6). We must look for a set of equations which (1) will satisfy this equation, (2) will reduce to (10–7) when $v \ll c$, and (3) which is as simple as possible. For simplicity let us try the linear transformations

$$x' = k(x - vt), \qquad t' = k(t - bx), \tag{10-8}$$

where k and b are independent of x and t and are to be determined. These equations satisfy conditions (1) and (2) above only if we make

$$k = \frac{1}{\sqrt{1 - v^2/c^2}} \quad \text{and} \quad b = \frac{v}{c^2}.$$

(The reader should verify that the conditions are satisfied when k and b have these values.) With these required values substituted in Eq. (10–8) we get

$$x' = \frac{x - vt}{\sqrt{1 - v^2/c^2}}, \qquad t' = \frac{t - vx/c^2}{\sqrt{1 - v^2/c^2}}, \tag{10-9}$$

which are the so-called *Lorentz space-time transformation equations* required by Einstein's postulate that the speed of light is the same in any inertial system. The most remarkable feature of these equations is the interrelationship between space and time in the two systems. The following two sections will illustrate this point.

10|6 THE APPARENT CONTRACTION IN LENGTH OF MOVING OBJECTS

Suppose that we have a rigid meter stick which when at rest in the laboratory is checked against a standard meter bar. Call the laboratory, or ground, the system S. For the system S', we shall take a very

---

* $\ll$ means "is much less than."

fast train which is moving in the x-direction with the speed v. Let the meter stick ride on the train with its length parallel to the direction of motion (see Fig. 10–5). An observer on the train checks the meter stick against a standard bar that he is carrying, and finds the length of the stick still to be one meter.

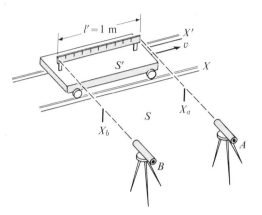

FIG. 10–5
A moving meter stick appears shortened to ground observers.

Observers on the ground set up telescopes along the tracks so that they can observe where, relative to S, the two ends of the stick are at some time t. The ground observers must first check their clocks with each other to be sure that they agree on the time t. They then say that they will observe the two ends of the moving meter stick at the *same* time t, or, as they say, "simultaneously." Observer A at $x = x_a$ finds the front end of the moving stick to be passing the point X_a on the ground at the same time that B at $x = x_b$ finds the other end passing the point X_b. Substituting in the Lorentz equation $x' = k(x - vt)$, we have for the conditions above

$$x'_a = k(x_a - vt), \qquad x'_b = k(x_b - vt),$$

where t *is the same* in each case. Subtraction yields

$$x'_a - x'_b = k(x_a - x_b),$$

or

$$x_a - x_b = \frac{1}{k}(x'_a - x'_b) = (x'_a - x'_b)\sqrt{1 - \frac{v^2}{c^2}}. \qquad \textbf{10–10}$$

Now $(x'_a - x'_b)$ is the length of the stick as checked by the observer on the train, and hence is exactly one meter. Then the length measured by the ground observers is *less* than one meter by the factor $\sqrt{1 - v^2/c^2}$.

Since motion is relative, the system S may be thought of as moving relative to S' with the velocity $- v$. Therefore we may turn the experiment around and have observers C, D, etc., on the train measure, *at their same time t'*, the length of an object on the ground. The result would be

$$x_a' - x_b' = (x_a - x_b)\sqrt{1 - v^2/c^2},$$

where the x's and x''s are not those in Eq. (10-10). We see that this apparent shortening effect is reciprocal.

The measured length of an object is less when the object and the observer are in relative motion than when they are at rest with respect to each other.

Thus we have derived the *Lorentz-Fitzgerald contraction* from Einstein's postulate.

10|7 THE APPARENT SLOWING OF MOVING CLOCKS

Let observer A at $x = x_a$ flash one signal at the time $t = t_a$ and another signal at $t = t_b$. The signals might be one minute apart by A's clock. Suppose that a group of observers in the moving system S' have synchronized their clocks and that the observer passing A at the time of the first flash looks at his clock and records the time t_a' of the flash. Another observer in S' similarly records the time t_b' of A's second flash as he passes A. From the Lorentz equations, we have

$$t_a' = k(t_a - vx_a/c^2), \qquad t_b' = k(t_b - vx_a/c^2),$$

where $x = x_a$ in each case. Subtraction yields

$$t_b' - t_a' = k(t_b - t_a) = \frac{t_b - t_a}{\sqrt{1 - v^2/c^2}}. \qquad\qquad \textbf{10-11}$$

Since k is greater than 1, the time between A's flashes is found to be longer by observers in the S'-system than it is according to A (who is in the system S). Conversely, due to the relativity of motion, the time between two events occurring at the same place in S' is found to be longer by observers in S than by observers in S'. We may sum up by saying that *a clock will be found to run more and more slowly the greater the relative motion between the clock and the observer.*

This effect becomes important in the case of high-energy cosmic particles of low mass, such as mesons, whose speeds relative to the ground may be only slightly less than that of light. These particles are observed to have short lifetimes, but with their high speeds they may travel considerable distances during their lifetimes. Since the speed of a meson gradually decreases as it loses energy in the earth's atmosphere, the velocity of the ground relative to the meson also changes. Hence for a given meson the lifetime usually referred to is that for a system in which the meson is at rest. Call this lifetime τ_0 and let τ represent the lifetime and $\bar{v}$ the average speed observed in the earth system, relative to which the meson is moving. Then $\bar{v}\tau$ is the distance traveled by the mesons according to earth observers. This distance has been measured. Experimenters have also determined τ_0 by creating artificially identical mesons that are practically at rest relative to the laboratory and then measuring their lifetime. We know that $\bar{v}$ cannot exceed c, the speed of light (see next section), yet the distance $\bar{v}\tau$ is found to be greater than $c\tau_0$. This means that τ must be greater than τ_0, or that it seems to us that time passes more slowly for the fast-moving meson than it does for us.

Example. A meson has a speed $v = 0.8c$ relative to the ground. Find how far the meson travels relative to the ground if its speed remains constant and the time of its flight, relative to the system in which it is at rest, is 2×10^{-8} sec.

Solution. Since $v = 0.8c$, $v^2/c^2 = 0.64$, $1 - v^2/c^2 = 0.36$. Relative to the earth the time of flight is

$$\tau = \frac{\tau_0}{\sqrt{1 - v^2/c^2}} = \frac{2 \times 10^{-8} \text{ sec}}{0.6} = 3.33 \times 10^{-8} \text{ sec.}$$

The apparent distance traveled is then

$$\Delta x = vt = 0.8c\tau$$
$$= (0.8 \times 3 \times 10^8 \text{ m/sec}) \times (3.33 \times 10^{-8} \text{ sec})$$
$$= 8.0 \text{ m.}$$

Note that this is greater than the distance the meson could travel in 2×10^{-8} sec even if its speed were that of light.

The same line of reasoning has been applied in a speculative way to the problem of interstellar space travel. How much we would like to know whether other stars have planetary systems and whether life exists on any of them! The nearest star is Alpha Centauri and it is about 4 light-years distant; the brightest star, Sirius, is about twice as far away. A *light-year* is the distance traveled by light in one year, which is about

6×10^{12} miles or 10^{16} meters. We shall see that, according to the relativity theory, we cannot hope to travel at speeds equal to, or greater than, the speed of light. Suppose, however, that we could launch a space ship at a speed $v = 0.8c$ relative to the earth and follow its progress through a telescope. We would find that it took $4/0.8 = 5$ years to reach the neighborhood of Alpha Centauri according to *our* clocks, but the people on the space ship would say that, according to *their* clocks, our 5 years was

$$5/\sqrt{1 - (0.8)^2} = 5/0.6 = 8\tfrac{1}{3} \text{ years,}$$

during which time they could travel well beyond Alpha Centauri. To reach Alpha Centauri the space travelers would only have to exist for $0.6 \times 5 = 3$ years of their lives in their ship. (Since moving clocks run slow, we would call these three years five of our years.) Another way of putting it is to say that to the moving space travelers the distance to Alpha Centauri would appear shortened from the 4 light-years measured by earth observers to $0.6 \times 4 = 2.4$ light-years, which could be covered in 3 years by a ship moving at $\tfrac{8}{10}$ the speed of light. There will thus be a double advantage in performing space voyages at speeds as close to that of light as possible.

Should a person travel to Alpha Centauri and right back at the speed $v = 0.8c$, he would arrive home 6 years older, while his friends on earth would have aged by 10 years! The reason that there is a reciprocity between observations made of a system S' from a system S and observations made of S from S', but not in the case of the returning space traveler and his earthbound friends, is that the space traveler *turned around* in order to return to earth. The space traveler did not stay in the same inertial system S', but the earth inhabitants did remain in their system S. This method of out-living your contemporaries is referred to as the *twin paradox*, since a twin who takes a fast trip should return home younger than his twin brother or sister who does not go away. As George Gamow has pointed out, if the speed of light were only 20 mi/hr, a postman bicycling around on his route all day every day might eventually be younger than a sedentary granddaughter!

If all of this sounds quite preposterous it is because it is unfamiliar. In our daily lives we do not encounter the length-shortening and time-slowing effects because at ordinary speeds these effects are far too small to measure. However, this does not mean that they do not exist; when dealing with high-speed particles for which the effects are measurable, we find that the predictions of relativity theory are true, so the theory must be accepted along with other successful ones. There is, however, great

245

doubt that its extension to distant space travel will ever be practicable according to the present state of our knowledge. Problems 10 and 11 at the end of the chapter will help to make this point clear.

10|8 THE ADDITION OF VELOCITIES

Another interesting theorem that may be derived from the relativity principle concerns the addition of velocities. We are accustomed to saying that if an airplane whose airspeed is 500 mi/hr is flying downwind in a wind whose velocity is 50 mi/hr, then the plane's ground speed must be 550 mi/hr. This is true for all practical purposes. However, when we come to add velocities which approach in magnitude the speed of light, we find that the above method of simple addition does not apply.

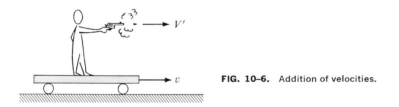

FIG. 10–6. Addition of velocities.

Suppose, for example, that a bullet is fired in the forward direction from a fast train whose ground speed is v (Fig. 10–6). The muzzle velocity of a bullet is measured relative to the gun and is determined by the powder in the shell. Call the muzzle velocity V' and let V be the speed of the bullet relative to the ground. Then the relativity theory (with the aid of considerable mathematics) predicts that

$$V = \frac{v + V'}{1 + vV'/c^2},$$ 10–12

where c is the speed of light. As a result of Eq. (10–12) it turns out that V cannot exceed c so long as V' and v are less than c.

Example. Given $v = 0.9c$ and $V' = 0.9c$; find V.

Solution. From Eq. (10–12), we find that

$$V = \frac{0.9c + 0.9c}{1 + 0.81} = 0.9945c,$$

whereas according to the Galilean transformations the answer would be $V = 1.8c$.

Since the relativistic composition of velocities makes it impossible to build up a velocity greater than c out of two velocities each less than c, we see that *the velocity of light in vacuo is the maximum attainable velocity for a material object.*

10|9 RELATIVISTIC DYNAMICS AND THE VARIATION OF MASS WITH VELOCITY

We turn now to the second part of Einstein's relativity principle, which states that the laws of physics are the same, or invariant, in any inertial system. In particular we shall consider the invariance of Newton's law of motion and Newton's action-reaction law.

Newton's principle, or law, of motion in its most general form states that the net force **F** acting on a body equals the time rate of change of the momentum of the body. Newton's principle, or law, of action and reaction states that when two bodies A and B interact, as in a collision, the action of A on B is equal and opposite to that of B on A. Together these two principles tell us that in a collision or explosion momentum must be conserved, or the total momentum of the system must remain constant.

A theoretical study of various kinds of collisions shows that if it is assumed that momentum is conserved in one inertial system S, then it is not always possible for momentum to be conserved in another inertial system S', moving relative to S, unless an additional assumption is made. This new assumption must be that the mass of an object appears to be greater the faster the object is moving relative to the observer.

Let m_0 be the mass of an object as measured when the object is practically at rest relative to an observer and let m be its apparent mass when moving with the speed v relative to the observer. Then if one assumes that Newton's laws are invariant with respect to different reference systems, one must postulate that

$$m = \frac{m_0}{\sqrt{1 - v^2/c^2}}. \qquad \textbf{10-13}$$

Here m_0 is often called the "rest mass" and m the "moving mass" of the object. Since this postulate is required if Newton's laws, together with the relativity principle, are to hold true, Eq. (10–13) is a derived law rather than an arbitrary hypothesis.

While the mass of a body is evidently a function of its speed relative to the observer, this dependence of m on v is not significant until v reaches values of over 1% that of light, that is, over 1800 mi/sec or 3×10^6 m/sec.

The variation of mass with speed is shown graphically in Fig. 10–7. It is really only in the atomic world that we encounter speeds of $0.5c$ and greater, but experiments with high-energy electrons in an x-ray tube, and with still more energetic electrons from radioactive materials, have enabled physicists to verify Eq. (10–13) for values of v in excess of $0.99c$.

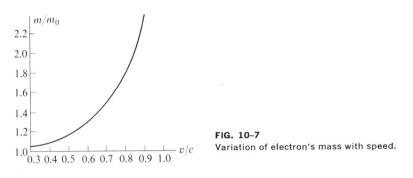

FIG. 10–7
Variation of electron's mass with speed.

Example. Find m/m_0 when $v = 0.99c$.

Solution. From Eq. (10–13) we have

$$\frac{m}{m_0} = 1/\sqrt{1 - (0.99)^2} = 1/\sqrt{0.02} = 7.09.$$

This represents an increase in mass of over 600% of the rest mass!

10|10 EINSTEIN'S MASS-ENERGY RELATIONSHIP

We saw in Chapter 5 that the kinetic energy (KE) of a body is calculated by computing the work required to accelerate the body from rest to its given state of motion at some speed v. Work is measured by the product of force and distance. If the mass of a body increases to a value greater than its rest mass, it will require a corresponding greater force to overcome the increased inertia of the body and maintain the same acceleration. Thus more work must be done than would be necessary if the mass did not increase with the speed. Therefore, the KE must be represented by a quantity that is greater than the $\frac{1}{2}mv^2$ computed in Chapter 5.

The actual amount of work performed by the increasing force needed to accelerate an object, whose rest mass is m_0, up to the speed v must be computed with the help of the integral calculus. This work, or the KE acquired by the object, is found to be

$$\text{KE} = \frac{m_0 c^2}{\sqrt{1 - v^2/c^2}} - m_0 c^2, \qquad\qquad \textbf{10–14}$$

or, if we use Eq. (10–13),

$$\text{KE} = mc^2 - m_0 c^2. \qquad\qquad \textbf{10–15}$$

These relativistic expressions for KE do not appear to bear much resemblance to that of nonrelativistic mechanics, but actually Eqs. (10–14) and (10–15) reduce to $\frac{1}{2}mv^2$ when v/c is sufficiently small. [Those familiar with the binomial theorem may apply it to the term $(1 - v^2/c^2)^{-1/2}$ in Eq. (10–14) and verify the above statement.] When v is large and approaches the magnitude c, we see that the first term in Eq. (10–14) approaches infinity. In other words, it would require an infinite amount of work to accelerate a material body up to the speed of light and hence this is impossible. Again we see that the speed of light is an upper limit for the velocities of material objects.

According to Eq. (10–15),

$$\text{KE} = (m - m_0)c^2 = \Delta m\, c^2, \qquad\qquad \textbf{10–16}$$

or the kinetic energy of a moving body equals its gain in mass times c^2. We may also say that the apparent mass of a body increases linearly with its kinetic energy, so that an increase in mass is an indication and measure of the gain in kinetic energy. It is also found that an increase in the potential energy of a system of particles is accompanied by a similar increase in mass equal to the gain in energy divided by c^2. Therefore we may say, in general, that *the gain (or loss) in the energy of a system is equal to the gain (or loss) in its apparent mass multiplied by c^2*.

We may go one step further and interpret the term $m_0 c^2$ in Eq. (10–15) as the *rest energy* of a body whose rest mass is m_0. This rest energy may be regarded as a form of internal energy inherent in the nature of the particles out of which matter is composed. If we solve Eq. (10–15) for mc^2, we then have

$$mc^2 = \text{rest energy} + \text{kinetic energy} = \text{total energy}.$$

If here we let E stand for the total energy, we arrive at Einstein's famous principle of the equivalence of mass and energy,

$$E = mc^2. \qquad\qquad \textbf{10–17}$$

The value of any theory is measured by its success in predicting new results. In this respect Einstein's theory has been outstanding. In nuclear physics the equivalence of mass and energy has been put to the test repeatedly and it has always been confirmed. With the ability to measure the masses of atomic particles to a high degree of accuracy, nuclear physicists have been able to predict the energy changes accompanying nuclear and particle transmutations, and they have also been able to verify their predictions experimentally. The whole subject of nuclear energy (popularly called "atomic energy") illustrates the usefulness of the above principle. The energy exchanges involved in chemical reactions must also be accompanied by corresponding mass changes, but in this relatively low-energy field the mass changes are too small to be detected experimentally. (See Chapter 14.)

While chemical reactions and most nuclear ones involve a rearrangement of atomic or subatomic particles, with a consequent change in their potential energy, there are also pair production and pair annihilation reactions, which are, incidentally, additional examples of the conservation of charge principle. In the latter reactions a particle of rest mass m_0 and charge e encounters and neutralizes a particle of rest mass m_0 and charge $-e$. One particle is said to be the *antiparticle* of the other; examples are the proton and antiproton, also the positron and the electron. When such particles come together, not only do their charges neutralize each other, but the particles cease to exist as such and in their place electromagnetic radiation is born. When the colliding particles have little KE it is found that the energy of the radiation produced just equals $2m_0c^2$. This has been interpreted in different ways. Some physicists say that in this process the mass $2m_0$ is converted into radiant energy. However, we shall see in the next chapter that it is fruitful to consider mass to be associated with radiation as well as with material particles. Radiation is found to possess inertial properties. Pair annihilation, and the inverse process of pair production (see following example), must then be interpreted as processes in which mass and energy are transformed together from one form to another. In general, it would seem preferable to regard mass as a property of energy, or energy as a property of mass, the two being inseparable. The basic conservation principle then is the *conservation of mass-energy*.

Example. *Pair production.* It was predicted theoretically by Dirac in 1928 and experimentally verified by Carl Anderson in 1932 that radiant energy might, under certain circumstances, be made to produce simultaneously an electron and a positron. The necessary conditions include:

(1) the radiant energy must be concentrated in units (photons) of energy E greater than $2m_0c^2$, m_0 being the mass of the electron and also that of the positron, (2) the process must occur in the presence of matter in order that some additional particle may be at hand to absorb the recoil momentum and thus ensure conservation of momentum. Given that

$$m_0 = 9.1 \times 10^{-31} \text{ kg},$$

what is the minimum value for E, the energy per photon of the radiation?

Solution. We must have

$$
\begin{aligned}
E = 2m_0c^2 &= 18.2 \times 10^{-31} \times 9 \times 10^{16} \text{ kg-m}^2/\text{sec}^2 \\
&= 163.8 \times 10^{-15} \text{ J} \\
&= \frac{163.8 \times 10^{-15} \text{ J}}{1.6 \times 10^{-19} \text{ J/eV}} \\
&= 1.02 \times 10^6 \text{ eV, or } 1.02 \text{ MeV (million electron-volts).}
\end{aligned}
$$

This figure has been verified experimentally.

PROBLEMS

1. Referring to the two swimmers in Section 10–4, take the current $v = 3$ ft/sec and the speed c of each swimmer in still water to be 5 ft/sec. Find

 a) the times t_1 and t_2 when $\overline{AC} = \overline{AB} = 100$ ft, and

 b) the distance $\overline{AC}$ for which, with $\overline{AB} = 100$ ft, the swimmers would have the same times.

2. A certain young lady decides on her twenty-fifth birthday that it is time to slenderize. She weighs 200 lb. She has heard that if she moves fast enough she will appear thinner to her stationary friends.

 a) How fast must she move to appear slenderized by a factor of 50%?

 b) At this speed what will her mass appear to be (to her stationary friends)?

 c) If she maintains her speed until the day she calls her twenty-ninth birthday, how old will her stationary friends claim she is according to their measurements?

3. Define the momentum of a body, according to relativistic dynamics.

4. For what kinetic energy in electron-volts (see Problem 9 of Chapter 7) will the ratio m/m_0 for an electron be

  a) 1.1,   b) 2,   c) 10?

5. Repeat Problem 4 for protons instead of electrons.

6. For what energy in electron-volts will the ratio v/c for an electron be

  a) 0.10,   b) 0.50,   c) 0.90,   d) 0.99?

7. A red-hot sphere of iron (specific heat capacity $= 0.11$ kcal*/kg-°C) weighs close to 1 kg. If the sphere cools 1200C°, what is the equivalent loss in mass?

8. Solar radiation reaches the earth, which is 90,000,000 mi or 1.5×10^{11} m away from the sun, at the rate of about 1.4×10^3 W/m². At what rate is the sun losing mass due to its radiation in all directions?

9. How long would a space traveler think he exists while traveling, relative to the earth, at a speed $v = 0.9c$ to a star that earthmen say is 40 light-years away?

10. At a constant acceleration of the magnitude of g, or 9.8 m/sec², how long would it take a space ship to reach the speed $v = 0.9c$? Compute your answer in days.

11. Suppose that in problem 9 the space traveler and his ship (without fuel) have a rest mass of 1000 kg.

  a) Compute the equivalent mass of the traveler and ship at a speed $v = 0.9c$.

  b) If the fuel itself did not have to be accelerated, what mass of fuel would be needed to furnish the required KE if 1% of the mass of the fuel could be passed to the space ship as energy? (At present man has been able to convert only a few tenths of 1% of the mass of nuclear fuel into mass associated with radiant, kinetic, and other forms of energy.)

  c) Show that if half of the fuel must, on the average, also be accelerated, then the space ship could not be made to approach a speed at which relativistic effects would be important.

---

* This is our large calorie, which equals 4180 J.

BARNETT, L.,	*The Universe and Dr. Einstein*, Harper, 1948.	**References**
EDDINGTON, A. S.,	*The Nature of the Physical World*, Macmillan, 1929.	
EINSTEIN, A.,	*The Meaning of Relativity*, Princeton University Press, 1950.	
EINSTEIN, A., AND L. INFELD,	*The Evolution of Physics*, Simon and Schuster, 1938.	
GAMOW, G.,	*Mr. Tompkins in Wonderland*, Cambridge University Press, 1957.	
INFELD, L.,	*Albert Einstein, His Work and Its Influence on Our World*, Scribner's, 1950.	
JEANS, J.,	*The New Background of Science*, Cambridge University Press, 1953.	

11|1 THE NEED FOR A NEW PRINCIPLE As has been said, the laws of classical physics apply very well to the behavior of objects that we can see with our eyes. However, in modern atomic physics we talk about and experiment with particles too small to be seen even under the highest powered microscope. Physicists have become so familiar with the properties of atoms and electrons that their existence seems just as real as that of a tennis ball. This sense of reality has been enhanced by the invention of models, such as Rutherford's planetary model of the atom, in which one pictures electrons circling the atomic nucleus just as the planets do the sun. Such models have helped us to assimilate and remember many different experimental facts, and they

The Quantum Principle

have led to the discovery of new properties possessed by atoms and particles. However, one must always guard against taking a model too seriously, as has sometimes been done in atomic physics. Physicists talked so much about the planetary model of the atom that some people began to think that given a more powerful microscope, one might actually be able to see the electrons circulating in definite orbits, just as we see the planets move in the sky; it was hard for such people to understand the later developments of atomic theory, according to which there is no sense in talking about an electron occupying a definite position at a given time or being in a sharply defined orbit. We are really on safe ground only when we speak of the *observable* properties of particles that we cannot see and it has become necessary to keep this point more and more in mind. Thus in this chapter we shall discuss measurable energy exchanges on the atomic level, but say little about hypothetical electron orbits.

Let us first list some of the observable facts that cannot be explained by the twelve fundamental laws of the preceding chapters, but require the introduction of a thirteenth law, the quantum principle. In historical sequence the most important of these facts are those relating to (1) the frequency distribution of the radiation emitted by incandescent solids, (2) the photoelectric effect, (3) atomic spectra, (4) the Compton effect, (5) the diffraction properties of corpuscular particles, and (6) the "leakage" of particles through a potential barrier, for example, an alpha particle escaping from an atomic nucleus. Of course there are many other important related facts, but those listed above are outstanding and at the same time so diverse in their nature as to illustrate the wide scope of the quantum principle. That a single unifying principle can "explain" all of these facts is a source of wonder; at the same time, their diversity suggests that it may be difficult to state the quantum principle as easily and concisely as we could our other laws. As a matter of fact, the full meaning of the quantum principle was realized only a little at a time; it was approached from different directions, just as explorers have gone into an unknown territory first from one side and then from another.

It would be nice if one could simply say that the quantum principle states that radiant energy consists of discrete units, just as matter and electricity do. The principle does say that some of the properties of light may be described in this way, but it says a lot more too, such as that all particles (corpuscular and light) have a wave-particle duality; that atomic systems can exist only in certain energy states; and that there is a basic uncertainty related to the simultaneous measurement of the position and momentum of a particle. In order to understand all of this it is necessary to study in more detail the facts listed above and how the quantum principle pertains to them. As we do this the reader will note that the unifying thread that runs through the explanation of all of these facts is the introduction of Planck's constant h, a new universal constant not needed in classical physics.

11|2 BLACKBODY RADIATION

When a solid is heated to incandescence it emits radiation of a continuous range of frequencies, such as we find in solar radiation. Just how the total energy is spread among the different parts of the spectrum is something that depends on the temperature and color of the emitting body. However, for all blackbodies the spectral distribution of the radiation is a function of the temperature only. A *blackbody* is defined as one that does not reflect any of the radiation falling on it. The most practical way of visualizing a perfect blackbody is to think of a cavity, like a furnace, with a very tiny hole through which one can see into the interior from outside (Fig. 11–1). Any outside radiation that falls on the hole will go in, be reflected by the inside walls, and eventually be absorbed by them. When equilibrium is reached at any given temperature, the walls of a furnace must be emitting and absorbing radiation at equal rates.

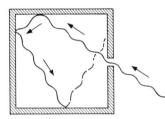

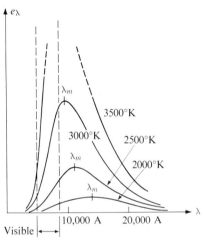

FIG. 11–1
A cavity with a small opening is equivalent to a blackbody.

FIG. 11–2
Distribution of the energy radiated by a blackbody at four different temperatures.

If one uses a grating, or prism, to form the spectrum of cavity radiation, and if one also measures the intensity contained in a given range of wavelengths, say a range of 100 Å (10^{-8} m), first in one part of the spectrum and then in another, one will observe the following facts: (1) Very long and very short wavelengths carry very little of the energy. (2) There is a certain wavelength λ_m at which the energy per given range of wavelength is a maximum. (3) As the temperature of the furnace increases, λ_m decreases in such a manner that the product $\lambda_m T$, where T is the Kelvin temperature, is a constant equal to 0.0029 m-°K. (4) As the temperature increases, the total energy emitted at all wavelengths increases with the *fourth power* of the Kelvin temperature. One may summarize all of these facts with the curves shown in Fig. 11–2, where e_λ, the rate at which energy is emitted per unit area per given range of wavelength, is plotted against the wavelength λ for several temperatures. While fact (4) above can be explained in terms of classical thermodynamics, the laws we have discussed so far do not explain the shape of the curves shown in Fig. 11–2, nor do they even account for the presence of the maximum at λ_m.

257

This was the situation in 1900 when the great German theoretician Max Planck (1858–1947) tackled the problem of finding a theory that would explain the above facts. Planck proceeded to follow a wise plan. First he studied the previous literature on the subject and found that two formulas, based on classical ideas, had been proposed, one describing one part of the energy-wavelength curves of Fig. 11–2, and the other describing another part of the same curves. Neither formula described the complete curves. Second, Planck put these two formulas together in such a manner that he obtained an empirical formula which fits the entire wavelength range. This formula is

$$e_\lambda = \frac{C_1}{\lambda^5(e^{C_2/\lambda T} - 1)},$$
11-1

where λ is the wavelength in the neighborhood of which the emission is e_λ per unit wavelength range, e is the base of natural logarithms, namely $2.718\ldots$, and C_1 and C_2 are constants to which Planck had to give values which would make Eq. (11–1) agree quantitatively with the experimental facts. Third, Planck studied the earlier theories to see in what respect they failed to explain his empirical formula. Fourth, Planck found that to obtain a successful theory he had to make a radical alteration in one of the previous theories. This alteration was such an arbitrary postulate that even Planck was at first reluctant to make it, but it is one that has successfully withstood the test of time. Let us see what this arbitrary postulate was.

Planck commenced his theory by speculating about how the walls of a furnace emit radiation. He assumed that these walls contain oscillators which vibrated with all possible frequencies. Hertz had shown that when electrons move back and forth in an antenna with the frequency f, radiation of this frequency is emitted (see Chapter 9), so Planck assumed that oscillators of the frequency f were responsible for giving out and absorbing this frequency. When the only force acting on an oscillating particle is a restoring force proportional to its displacement from equilibrium, the oscillator will vibrate with the same frequency regardless of the amplitude of its vibration; the amplitude only determines the total energy of the oscillator. In other words, an oscillator, while emitting energy, must diminish its amplitude; while absorbing energy, it must increase its amplitude.

In the case of such familiar oscillators as a pendulum and a vibrating tuning fork, it appears that the energy may decrease *continuously*. Planck found that he could not assume this for his cavity oscillators if his theory was to lead to the correct empirical radiation law. Instead, he had to postulate that

an oscillator of frequency f can exist only in states for which the energy E is a whole number multiple n of the product hf,

where h is a constant that is the same for oscillators of all frequencies, i.e.,

$$E = nhf, \qquad n = 0, 1, 2, 3, \ldots \qquad \text{11-2}$$

Planck found that if he did not restrict n to integral values, but allowed it to vary continuously, then his subsequent theory led to one of the incorrect formulas for e_λ.

If energy is plotted vertically upward, the *energy states* of an oscillator of frequency f that are allowed according to Eq. (11–2) may be represented as a series of equally spaced lines, like the rungs in a ladder, as shown in Fig. 11–3.

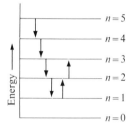

FIG. 11–3. Energy-level diagram for a Planck oscillator. The energy E corresponding to any level is *nhf*.

To his first postulate Planck had to add another equally arbitrary one, namely that

an oscillator can emit energy only by going from one energy state to the next lower one,

and that it can absorb energy only by going up one step in the energy-state ladder. This implies that a cavity oscillator does *not* emit energy while in a given state. Now, Planck's oscillators were presumably electrons moving to and fro, and we have seen that such accelerated charges should, according to the laws of electromagnetism, continually radiate energy in the form of electromagnetic waves. Planck simply assumed that his cavity oscillators did not radiate as do free electrons in a wire.

With the above assumptions Planck derived the following formula for e_λ:

$$e_\lambda = \frac{2\pi hc^2}{\lambda^5(e^{hc/\lambda kT} - 1)},\qquad \text{11-3}$$

where h is Planck's new constant, c is the speed of light, and k is Boltzmann's constant. This formula agrees with all of the facts listed earlier in this section; furthermore, it is of the same form as Planck's empirical law, Eq. (11–1).

The agreement in form of Eq. (11–3) and Eq. (11–1) assured Planck's theory of qualitative success. More surprising was its quantitative agreement with the facts. By comparing the two equations we see that $2\pi hc^2$ should equal C_1. Since h is a new constant, Planck could choose its value so as to make $2\pi hc^2 = C_1$. The required value of h came out to be

$$h = 6.6 \times 10^{-34} \text{ J-sec.} \qquad\qquad \textbf{11–4}$$

Further comparison of the two equations for e_λ shows that hc/k should equal C_2. Since Planck had already chosen a value for h, and since the values of c and k had also been determined, he could test his theory by computing hc/k and comparing its value with the experimental value of C_2, namely

$$C_2 = 0.014 \text{ m-}^\circ\text{K.}$$

If we apply the above test, we find that

$$hc/k = \frac{(6.6 \times 10^{-34} \text{ J-sec}) \times (3 \times 10^8 \text{ m/sec})}{1.38 \times 10^{-23} \text{ J/}^\circ\text{K}}$$

$$= 0.0143 \text{ m-}^\circ\text{K.}$$

This equality of hc/k and C_2 furnishes strong experimental support to Planck's theory.

In spite of its qualitative and quantitative success, Planck's theory was not universally accepted at first because its postulates were so radically new. For example, some said that Planck's theory was *built* to explain cavity radiation and so of course it *did* explain cavity radiation, *but nothing more*. Time has shown the incorrectness of this argument. Instead we may now say that very little in atomic physics can be explained without the introduction of a constant whose magnitude and units are the same as those of h in Eq. (11–4). Some people still find it difficult to accept a point of view at odds with that of everyday life, even though this new way of thinking is *not* applied to the events of everyday life, but to ones of an entirely different order of magnitude. Wise scientists must, on the one hand, keep their minds open to new ideas, while on the other hand they must use judgment in accepting them.

Example. A pendulum has a period of 2 sec. The bob, whose mass is 50 gm, is pulled out sufficiently far to raise it 10 cm above its equilibrium position. (a) Compute the initial energy of the bob. (b) If the bob loses the energy hf, what fraction of its original energy will be lost?

Solution

a) Here $T = 2$ sec, $f = 1/T = 0.5$/sec. The bob's initial energy E is potential, therefore

$E = mg \times$ height lifted

$\quad = (5 \times 10^{-2}\,\text{kg}) \times (9.8\,\text{m/sec}^2) \times (0.1\,\text{m})$

$\quad = 4.9 \times 10^{-2}\,\text{J}.$

b) Here $hf = (6.6 \times 10^{-34}\,\text{J-sec}) \times (0.5/\text{sec})$

$\qquad\quad = 3.3 \times 10^{-34}\,\text{J}.$

$hf/E = 3.3 \times 10^{-34}/4.9 \times 10^{-2} = 6.7 \times 10^{-33}.$

Thus in this case hf represents a change in the energy that is far too small to be measured. This shows that the postulate of discrete, or individually separated, energy states is not in conflict with the observation that in everyday life we do not observe such a limitation on the energy of a mechanical system.

11|3 THE PHOTOELECTRIC EFFECT

While Hertz was experimenting on the production of electromagnetic waves by means of a spark discharge between two neighboring spheres, he observed that ultraviolet radiation produced in the discharge caused negative electric charges to be emitted from the negative sphere. These negative charges were later identified as electrons and it was also found that they could be ejected from the cold surface of an alkali metal, such as sodium, by visible light. This phenomenon is called the *photoelectric effect*. We are all familiar with its application to light meters in photography and to "magic eye" devices for opening doors, etc.

Further study of the photoelectric effect disclosed the following facts.

a) If the incident light is monochromatic (of one frequency only), then the number of photoelectrons emitted is proportional to the *intensity* of the light.

b) Increasing the *intensity* of the light in (a) does not affect the KE with which the individual electrons are emitted.

c) If a certain light causes photoelectrons to be emitted, then, when the light is turned on, emission commences immediately, with no apparent delay, even when the light is very weak.

d) For a given metal the KE of the individual photoelectrons increases with the *frequency* of the light used. There is a certain threshold frequency f_0 below which no emission occurs. For frequencies above f_0, the KE of the photoelectrons is proportional to $(f - f_0)$, where f is the frequency of the light used.

Here, then, are four sets of facts that a photoelectric theory must explain. Let us take the theory that light consists of waves and test it by determining how well it accounts for these facts. The wave theory explains fact (a) satisfactorily, since with light of greater intensity more energy must be absorbed by the metal and more electrons could thus be emitted; but, according to this reasoning, we would also expect some of the electrons to come out with greater individual KE when the light is more intense, and this prediction is contrary to fact (b). Fact (c) is also hard to explain. Some work must be required (call it W) to get an electron out through the surface of a metal, or else electrons would "evaporate" from the metal all the time. How can an electron gather this energy to itself from a succession of broad waves in which the energy per electron cross section is very small? Calculations show that if an electron could only absorb the energy falling on it directly, it would in some instances have to wait many seconds in order to accumulate sufficient energy to escape. The wave theory fails completely to explain fact (d) above, so altogether its score of success is very low.

In 1905 Einstein proposed another theory, one suggested by Planck's work on blackbody radiation. If, argued Einstein, radiation of frequency f is emitted in units of energy equal to hf and if it is absorbed by the oscillators of the cavity walls in units of the same energy, then why should we not regard the energy of an electromagnetic wave of frequency f as being concentrated in small bundles, each of energy hf? These bundles of energy are called *photons*, or *light quanta*. According to this theory, an increase in the *intensity* of a beam of light implies an increase in the *number* of photons passing through unit cross-sectional area of the beam in unit time, while an increase in the *frequency* of the light corresponds to an increase in the amount of *energy* carried *per photon*. As an analogy, consider two machine guns, A and B, and let B fire bullets of higher calibre than A. The bullets correspond to the photons in a beam of light. If either gun is caused to fire more bullets per second, the total energy output, or intensity of the firing, will increase, but since the energy of each bullet depends only on the powder in its case, the energy of the individual bullets will not be affected by the rate of firing. On the other hand, the bullets from gun B will possess more energy than those from gun A.

Shifting from gun A to gun B corresponds in the case of light to increasing the frequency of the light.

Applying the photon theory to the photoelectric effect, Einstein reasoned as follows: If light of frequency f falls on the surface of a metal, then concentrated bundles of energy, each of magnitude hf, must strike the metal and penetrate to where they are stopped by the electrons in the metal. Thus a photon of energy hf may give all of its energy to one electron and the latter may use this energy to escape through the surface of the metal. The work function W represents the portion of the energy given to the electron that must be used in getting the electron out of the metal. What is left over should, according to the conservation of energy principle, be the KE of the emitted electron. The above reasoning may be expressed mathematically by the equation

$$hf - W = \text{KE of photoelectron,} \qquad\qquad \textbf{11-5}$$

called *Einstein's photoelectric equation;* W varies with the metal used, but not with the frequency of the light. The threshold frequency f_0 is that for which the electrons escape with negligible KE, or

$$hf_0 = W. \qquad\qquad \textbf{11-6}$$

A good analogy to the photoelectric effect is that of a group of prisoners in jail, each under a bond W. A friend arrives, goes at random to one of the group, and offers to loan him x dollars. If x is less than W, the loan is declined; accepting the money will not enable the prisoner to get out of jail and he cannot spend money in jail. However, if x is greater than the bond W, the prisoner eagerly accepts the x dollars, pays his bond and leaves the jail with $(x - W)$ dollars in his pocket, which he may spend at the tavern of his choice!

Example 1. Light of wavelength $\lambda = 6200 \,\text{Å}$ falls on a metal and photoelectrons are ejected, each with a KE $= 0.3 \,\text{eV}$. Find (a) the energy in electron-volts of each photon of the light, and (b) the work function of the metal in electron-volts and joules.

Solution. The energy per photon is hf. Since $\lambda f = c$, the speed of light, we have

$$hf = \frac{hc}{\lambda} = \frac{(6.6 \times 10^{-34}\,\text{J-sec}) \times (3 \times 10^{8}\,\text{m/sec})}{6200 \times 10^{-10}\,\text{m}}$$

$$= 3.2 \times 10^{-19}\,\text{J}.$$

Recall that $1\,\text{eV} = 1.6 \times 10^{-19}\,\text{C} \times 1\,\text{V} = 1.6 \times 10^{-19}\,\text{C} \times 1\,\text{J/C} = 1.6 \times 10^{-19}\,\text{J}$, so that

$$hf = \frac{3.2 \times 10^{-19}\,\text{J}}{1.6 \times 10^{-19}\,\text{J/eV}} = 2\,\text{eV}.$$

Substitution in Eq. (11–5) gives us

$$2\,\text{eV} - W = 0.3\,\text{eV}, \quad W = 1.7\,\text{eV} = 2.72 \times 10^{-19}\,\text{J}.$$

Example 2

a) Suppose that light of 3100 Å strikes the same metal as in the previous example. With what KE will the photoelectrons now escape from the metal?

b) What is the longest wavelength (λ_0) that will give photoemission from this metal?

Solution

a) Halving the wavelength of the light from 6200 Å to 3100 Å is equivalent to doubling the frequency, which in turn means doubling the energy per photon. Hence we now have photons of 4 eV energy. Since W is unchanged in value, $W = 1.7\,\text{eV}$. The photoelectric equation now tells us that

$$4\,\text{eV} - 1.7\,\text{eV} = 2.3\,\text{eV} = \text{KE of photoelectrons}.$$

b) Since $hf_0 = W = 1.7\,\text{eV}$, while $hf = 4\,\text{eV}$ for $\lambda = 3100$ Å, we have by proportion

$$\frac{\lambda_0}{\lambda} = \frac{c/f_0}{c/f} = \frac{f}{f_0} = \frac{hf}{hf_0} = \frac{4}{1.7} = 2.35,$$

$$\lambda_0 = 2.35 \times 3100\,\text{Å} = 7300\,\text{Å},$$

which corresponds to light in the visible red.

   The photoelectric effect furnishes one of the best methods for measuring Planck's constant h (see problem 6 and Experiment 18 at end of chapter). The value of h found from the photoelectric effect agrees, within the limits of experimental error, with the value it must be given to fit the blackbody radiation curve. This agreement constitutes further support for the postulates, made by Planck and Einstein, involving the new constant h.

11|4 ATOMIC SPECTRA AND ENERGY STATES

For about 200 years it has been known that an incandescent gas emits light whose spectrum consists of a series of bright lines, as in Fig. 11–4(a). This *line emission spectrum* may also be obtained by passing an electric discharge through a tube containing the gas at low pressure. When the gas consists of a single chemical element, such as hydrogen, helium, or sodium, the pattern of lines or wavelengths is characteristic of the element. These patterns have been classified and used to identify the elements present in some unknown substance heated to incandescence.

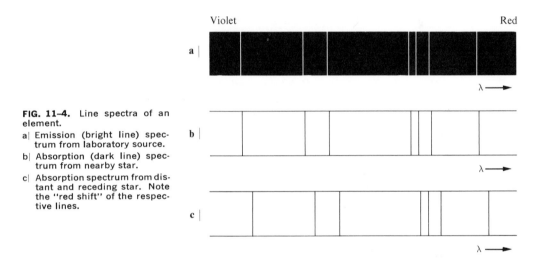

FIG. 11–4. Line spectra of an element.

a| Emission (bright line) spectrum from laboratory source.

b| Absorption (dark line) spectrum from nearby star.

c| Absorption spectrum from distant and receding star. Note the "red shift" of the respective lines.

In the case of the sun, the very hot interior emits a continuous spectrum of all wavelengths. As this white light passes through the cooler outer atmosphere certain wavelengths are absorbed from the continuous spectrum, leaving a continuous spectrum with dark lines superimposed upon it, as in Fig. 11–4(b). Generally speaking, the dark *line absorption spectra* obtained in this manner and in simulated laboratory experiments are pretty much the same as the line emission spectra of the absorbing elements. Thus line absorption spectra are useful in determining stellar compositions and the relative abundance of the chemical elements throughout the universe. Furthermore, the fact that the line spectra of distant stars show the same line patterns of certain elements as do the nearer stars, but with every line displaced to a lower frequency or longer wavelength (Fig. 11–4c), is taken as an indication that the distant stars are moving away from us. The decrease in frequency is assumed to be of the same nature as the drop in pitch of an automobile horn or a train whistle when the vehicle is moving away from the observer, a phenomenon in wave motion called the *Doppler effect*.

Let us now consider the question of how atoms of a given element produce line spectra. This problem was first successfully solved for the case of hydrogen atoms by the great Danish physicist Niels Bohr (1885–1962). Bohr tried to explain the spectrum of hydrogen because its atoms are the simplest of all. He adopted Rutherford's planetary model of the atom, which in the case of hydrogen reduces to one electron orbiting around a nucleus containing just one proton. As a start Bohr postulated that Coulomb's law and Newton's law of motion hold for such an atomic system. This was a pure guess, for these laws had never been tested on the atomic scale and so it was quite possible that they would be found not to hold inside atoms. Bohr's second postulate was equivalent to that which Planck made in regard to the emission of radiation by the oscillators in the walls of a cavity. Bohr assumed that

atoms may exist in states of definite energies without radiating, but transitions between these states may occur, and when an atom passes from an initial allowed state of energy E_i to a final one of lower energy E_f, a photon of energy hf is radiated such that

$$hf = E_i - E_f. \hspace{3cm} \textbf{11-7}$$

Note that Eq. (11–7) combines Einstein's definition of the photon, or quantum, with the principle of the conservation of energy; the energy lost by an atom is radiated as a single quantum. Continuing radiation results from the successive emission in a short time of billions of quanta by billions of atoms each undergoing similar transitions.

The essentially new feature in Bohr's theory was the fact that he had to postulate that *a nonradiating atom can exist in only one of a discrete set of energy states* and that these states are *not* equally spaced. Bohr found an arbitrary rule for determining the allowed states for a hydrogen atom. This rule will not be stated, since it has been superseded by a more general one, applicable to all atomic systems, which will be given later. However, we shall see that Bohr's theory, like Planck's, met with both qualitative and quantitative success.

The spectrum of hydrogen in the visible and nearby ultraviolet region consists of a series of lines that are closer and closer together (and also less intense) at the high-frequency end of the series, as shown in Fig. 11–5. Such a series suggests some sort of regularity, which implies that there must be an empirical formula, with a limited number of constants, that describes the above series. This formula had indeed been discovered long before Bohr's work by a Swiss schoolteacher, J. J. Balmer (1825–1898). Balmer's formula may be written as

$$\frac{1}{\lambda} = R_H \left(\frac{1}{2^2} - \frac{1}{n^2} \right), \qquad n = 3, 4, 5, \ldots, \qquad \textbf{11-8}$$

where λ is the wavelength of one of the lines of the series corresponding to an integral value of n greater than 2, and R_H is a constant whose experimental value is 10,967,758/m. (The high accuracy implied here is characteristic of wavelength measurements in modern optics.)

Bohr's theory led him to the formula

$$\frac{1}{\lambda} = \frac{2\pi^2 k_e^2 m e^4}{h^3 c} \left(\frac{1}{n_f^2} - \frac{1}{n_i^2} \right), \qquad \textbf{11-9}$$

where n_f is the value of the integer n for the final state of a transition and n_i its value for the initial state. In atomic physics a quantity, such as n, which is limited to integral or half-integral values, is called a *quantum number*.

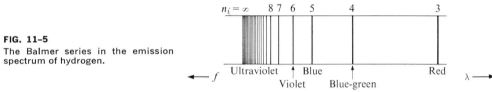

FIG. 11–5
The Balmer series in the emission spectrum of hydrogen.

To obtain a formula similar to Balmer's, Bohr put $n_f = 2$ and $n_i = 3, 4, 5, \ldots$

Equations (11–8) and (11–9) agree in form, but do they agree quantitatively? The final test of Bohr's theory was the comparison of his factor $2\pi^2 k_e^2 m e^4 / h^3 c$, which involves only previously measured constants, with the experimental value of Balmer's constant R_H. Actually

$$\frac{2\pi^2 k_e^2 m e^4}{h^3 c} = 1.097 \times 10^7 / \text{m}$$

within experimental errors, so that the agreement is most striking.

Bohr's theory of the hydrogen atom scored many other notable successes. For instance, if the value of n_f is put equal to 1 and $n_i = 2, 3, 4, \ldots$, one obtains a formula giving the correct wavelengths of another series of lines in the hydrogen spectrum, a series found in the far ultraviolet called the Lyman series. With $n_f = 3, n_i = 4, 5, 6, \ldots,$ one obtains an infrared series, and so on. The energy-level diagram for hydrogen is shown in Fig. 11–6; it furnishes a good method of illustrating and summarizing the known facts about the energy states and emission lines of hydrogen. The directed lines represent the transitions between states which give the various lines in the spectrum. Absorption of radiation by an atom corresponds to the reverse of emission and would be represented by lines directed upward from one energy state to a higher one.

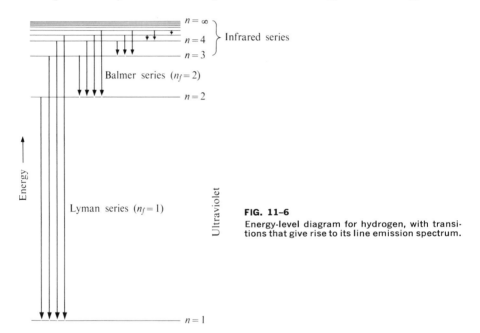

FIG. 11–6
Energy-level diagram for hydrogen, with transitions that give rise to its line emission spectrum.

Let us now consider the atoms of some other element. The line spectrum of an element indicates the existence of discrete energy states. The lowest of these states must be that of greatest stability and the one usually occupied; hence it is called the *normal state*. Above this state are the *excited* states of the neutral atom; an atom may be raised to one of these states by giving it energy, say through intense thermal agitation at high temperatures, through electron bombardment, or by the absorption of a photon of exactly the right amount of energy. Additional energy states may have to be postulated in order to account for all the lines of the emission spectrum. An energy-level diagram may then be drawn and possible transitions indicated. For all transitions

$hf = E_i - E_f.$

Example 1. Suppose that atoms of a certain element are found by electron bombardment to have an excited state 1.7 eV above the normal state. Also assume that the line spectrum of the element contains the following three wavelengths: 7300 Å, 5400 Å, and 3100 Å. Find the energy-level diagram with the fewest energy levels that will account for the facts.

Solution. In Fig. 11–7 the normal state is represented by the line A. There must be an excited state B at 1.7 eV above A. The transition from B to A accounts for the 7300-Å line in the spectrum (see Example 2 of Section 11–3).

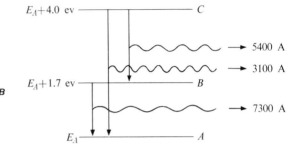

$E_A+4.0$ ev ———————— C

→ 5400 A

→ 3100 A

$E_A+1.7$ ev ———————— B

→ 7300 A

E_A ———————— A

FIG. 11–7
Atom with normal state *A* and excited states *B* and *C*.

To account for the 3100-Å line we must introduce a second excited state C that is 4 eV above A, since radiation of 3100 Å by an atom means emission of a 4-eV photon (see Example 2 of Section 11–3).

For the transition from C to B we have

$$hf = hc/\lambda = E_i - E_f, = (4.0 - 1.7) \text{ eV}.$$

$$\lambda = \frac{hc}{E_i - E_f} = \frac{(6.6 \times 10^{-34} \text{ J-sec}) \times (3 \times 10^8 \text{ m/sec})}{(2.3 \text{ eV}) \times (1.6 \times 10^{-19} \text{ J/eV})}$$

$$= 5.4 \times 10^{-7} \text{ m, or } 5400 \text{ Å}.$$

This transition thus accounts for the third line of the spectrum and no additional energy levels need be assumed.

Example 2. *The laser.* Suppose that the atoms of a given material have three allowed states A, B, and C, as in Fig. 11–7. Normally most of the atoms will be in state A, but if the material is exposed to radiation of frequency f such that $hf = E_C - E_A$, then photons of this frequency will be absorbed by atoms in state A, raising these atoms to state C. Some of the atoms in state C will return to state A and some will drop back to state B, where, after some delay, they may eventually return to state A. If the rate at which atoms are "pumped" up to state C and fall back to B is made to exceed the rate at which they fall from level B to level A, one will succeed in increasing the population of level B. This is made easier by the fact that impurities present may slow down the "leakage" from B to A. **269**

The overpopulation of state B may be relieved by passage through the material of a photon whose energy exactly equals $E_B - E_A$. Such a photon may trigger the almost simultaneous transition of many atoms from B to A, accompanied by the emission of identical photons. In this case radiation of an individual atom is affected by that of its neighbors in such a way that a coherent beam of light of high intensity is emitted in a certain direction, that is, the photons seem to combine to produce a single electromagnetic wave of large amplitude. This principle, discovered by Dr. Charles H. Townes at Columbia University, is that of the *laser*, i.e., *light amplification by stimulated emission of radiation*.

11|5 THE DUAL NATURE OF LIGHT

In discussing light, we may say that the three basic processes are *emission*, *propagation*, and *absorption*. In emission and absorption we are concerned with how light *energy* is produced at the expense of some other form of energy, and vice versa. Actually, we do not detect the presence of light until it falls on some absorbing object such as the human eye, a strip of film, or a photocell, in which cases the light energy is transformed into an electric nerve pulse, chemical energy, and the energy of photoelectrons, respectively. In this way the energy may be measured.

We have seen that photoelectric and other measurements indicate that light energy is emitted and absorbed in quantum units, rather than continuously. Remember, however, that emission and absorption processes have nothing to do with determining what path the light will follow as it travels from its source to its sink. Suppose that we use a system of slits to limit the path of the light and that we observe to where on a screen the greatest amount of energy travels. Then we can explain or predict these results correctly by means of the wave theory. In wave theory, we must introduce the concepts of wavelength λ and velocity of propagation c, both of which may be determined from experiments involving the propagation of light. The frequency, f, associated with light of wavelength λ is found from the wave relation $f = c/\lambda$.

We see, then, that the photon theory is used to describe emission and absorption, and the wave theory accounts for the propagation of light. These theories complement each other, since they apply to different phenomena and describe different aspects of light; by using each at the appropriate occasion our description of nature is more complete. On the other hand, we must not use both theories at the *same* time because they lead to contradictory conclusions. Evidently neither theory by itself is a completely right or wrong description of nature.

To remain on firm ground we must return to the discussion of what we measure. In diffraction experiments we measure the intensity, or energy per unit area per unit time, reaching various points on a screen. We may say that what we measure is the fraction of a large number of emitted photons received by a given area per unit time, or *the probability that one* photon will reach this area. The wave theory enables us to compute the correct probability. As far as individual particles are concerned, the completely causal relationship between past and future events which is assumed in classical theory is replaced, owing to insufficient information, by a statistical relationship giving the probability that a certain future event will occur. An analogous situation occurs in the throwing of dice (see Chapter 1).

11|6 THE COMPTON EFFECT

The photon theory has been further developed in order to explain other experimental results involving the interaction of light with matter. If a photon possesses energy, we may, if we wish, speak of the mass associated with, or equivalent to, that energy. We shall call this "associated mass" of the photon m. Note, however, that a photon does not have a rest mass, since there is no such thing as a photon at rest. A photon either moves with the speed of light, or it has been absorbed and no longer exists.

Let us combine Einstein's expression $E = hf$ for the energy of a photon with the one $E = mc^2$, which he derived from the relativity principle. We see that the "associated mass" of a photon must be

$$m = E/c^2 = hf/c^2. \qquad\qquad \textbf{11–10}$$

Since a photon travels with the speed of light, it must possess a momentum μ equal to the product of its mass and velocity, or of magnitude

$$\mu = mc = hf/c = h/\lambda. \qquad\qquad \textbf{11–11}$$

This relationship for the momentum of a photon is experimentally supported by the *Compton effect*.

When high-frequency radiation, such as a beam of x-rays, strikes matter the radiation is scattered in different directions. It was observed that some of the scattered x-rays were less penetrating and presumably had a longer wavelength than the incident radiation. In the early 1920's the American physicist Arthur H. Compton thoroughly investigated this effect and showed that his results could be interpreted only in terms of the above theory, in which a momentum as well as energy are ascribed to photons. Compton used x-rays of a single frequency f so that all of the incident photons would have the same energy hf and momentum hf/c. He measured the wavelength λ of the incident radiation and that of the scattered radiation, using the planes of atoms in a crystal for his grating. It was found that radiation scattered at a given angle ϕ contained a single new wavelength λ' that was longer than λ by a definite amount, independent of the incident wavelength used. Compton's explanation was based on the theory that a photon of the incident radiation may collide with an electron in the scattering material, causing the electron to recoil with some KE, thus leaving the quantum with less energy than it started with. The recoil electrons were soon observed and their energies were also measured.

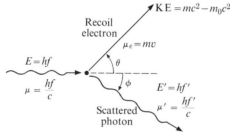

FIG. 11-8

Compton collision between photon and electron.

As in Fig. 11–8, consider a photon of energy hf to strike an electron and be scattered with the reduced energy hf'. From the principle of conservation of energy, we have

$$hf = \text{KE given to electron} + hf'. \qquad \textbf{11-12}$$

The relativistic expression for kinetic energy must be used, because for the recoil electrons the speed v may be an appreciable fraction of c, the speed of light.

Compton also applied the conservation of momentum principle to the collision, using Eq. (11–11) for the momentum of a photon. He computed the shift in wavelength for light scattered at an angle ϕ to be

$$\lambda' - \lambda = \frac{h}{m_0 c} (1 - \cos \phi). \qquad \textbf{11-13}$$

The factor $h/m_0 c$, called the *Compton wavelength*, has a numerical value of about 0.024 Å, where $1 \text{ Å} = 10^{-10}$ m. For monochromatic x-rays scattered through 90°, it is found that part of the beam undergoes just this change in wavelength.

Example. Suppose that $\lambda = 0.048$ Å and $\phi = 90°$ (or $\cos \phi = 0$). Then $\lambda' - 0.048$ Å $= 0.024$ Å, or $\lambda' = 0.072$ Å $= \frac{3}{2}\lambda$. Hence $f' = \frac{2}{3}f$ and the scattered photon retains two thirds of the original photon's energy, the other one third going to the recoil electron. The energy of a photon whose wavelength is 0.048 Å is a little over $250,000$ eV. Such photons may be produced by an x-ray tube operating at a potential above $250,000$ V (see problem 9).

11|7 THE DUAL NATURE OF ATOMIC PARTICLES

We must now face the problem of why atomic systems can exist in certain discrete energy states and not in any state for which the energy lies between the values allowed. In other words, why do quantum numbers, such as n in Bohr's theory, which are limited to integral values (half-integral in some cases) appear in expressions relating to atomic systems? This question was answered by extending to material particles the properties of waves discussed in Chapter 6; it was then postulated that such wave properties based on Huygens' principle, rather than the properties of a particle moving according to Newtonian mechanics, should be used to describe the behavior of atomic systems. This new theory is called *wave mechanics.*

Prince Louis de Broglie first proposed in 1924, in his thesis for the doctorate, that a material particle such as an electron might have a dual nature, just as light does. In the study of light the wave properties, involving λ, f, and c, are the more familiar ones. In terms of these properties we have defined the mechanical properties of the light corpuscles as follows:

$$E = hf, \qquad \mu = h/\lambda. \tag{11-14}$$

For a material particle it is the other way around; we are used to speaking of its mass m, momentum μ, and energy E. These are measurable quantities. Therefore when de Broglie postulated that such a particle may also have wave properties, he defined the associated wavelength λ and frequency f in terms of m, $\mu(=mv)$, and E. In doing so, de Broglie was guided by the relations for light [Eqs. (11–14)], and so he postulated that associated with a particle are waves of wavelength λ and frequency f, given by

$$\lambda = h/\mu, \qquad f = E/h. \tag{11-15}$$

The waves associated with a particle are not to be regarded as being mechanical or electromagnetic, but rather "probability waves." By this we mean that the intensity of the waves at any point will be taken as giving the fraction of a large number of similar particles, emitted with the same initial velocity, that will reach a given area in unit time, or the probability of one particle reaching that area. The waves are thus a device for computing the probability that a particle will behave in a certain way.

The behavior of a material particle is usually computed from the laws of mechanics. Is there, then, any need for a new method? If so, what are its advantages? De Broglie suggested that wave mechanics might stand in relation to particle mechanics as wave optics does to geometrical optics. In geometrical optics light is assumed to travel through a homogeneous medium in straight lines or rays. To explain the details of the diffraction pattern caused by an obstacle, one must resort to wave optics and be concerned with distances of the order of magnitude of one wavelength. When we compute the wavelength h/mv associated with an ordinary object, it turns out to be insignificantly small, but for an electron the mass m is very much less than for ordinary objects and h/mv is found (see example below) to be comparable to x-ray wavelengths. Since x-rays are diffracted by crystals, why should not such electrons be, too? This question was answered by Davisson and Germer who, while studying the scattering of electrons from metals, found that electrons are indeed diffracted by crystals just as x-rays are, and in accordance with the relation $\lambda = h/mv$. The wave properties of electrons are utilized in the electron microscope.

De Broglie also realized that in dealing with atomic systems, where the dimensions are also comparable to x-ray wavelengths, wave mechanics might be able to explain things that ordinary mechanics could not. For example, according to the Bohr theory of the hydrogen atom, only those electron orbits are allowed for which

$$mvr = n(h/2\pi), \qquad n = 1, 2, 3, \ldots, \qquad \textbf{11-16}$$

where mv is the momentum μ of the electron and r is the radius of its orbit about the nucleus of a hydrogen atom. This whole number rule was empirical when it was first proposed, the argument in its favor being that it worked. De Broglie suggested, by way of explanation, that if waves accompany an electron around its orbit, then a sort of resonance might occur when the circumference of the orbit equals an integral multiple of the electron's wavelength, as in Fig. 11–9. This idea, which was later refined by Schrödinger, amounted to saying that the allowed quantum orbits correspond to the various possible modes of vibration, or standing waves, in a string. For such a string, waves are reflected back and forth with constructive interference only when the wavelength λ is such that the complete distance down the string and back is an integral number n of wavelengths. Here the restriction that n must be a whole number has a physical basis, hence it was appealing to carry the idea over to the case of the atomic orbit and its electron waves.

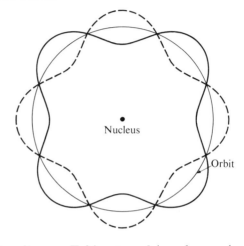

FIG. 11-9
Atomic orbit whose circumference equals four de Broglie wavelengths corresponds to the $n = 4$ mode of vibration.

Consider a circular orbit of radius r. Taking $\lambda = h/\mu$, where μ is linear momentum, de Broglie put the orbital circumference $2\pi r$ equal to $n\lambda$, or

$$2\pi r = n\lambda = n(h/\mu) = n(h/mv), \qquad mvr = n(h/2\pi),$$

which is the same as Eq. (11-16).

Bohr viewed the transition of a hydrogen atom from one state to another as being equivalent to the jump of its electron from one allowed orbit to another such orbit. According to wave mechanics such an atomic transition corresponds to the fading out of the waves accompanying one mode of vibration and the appearance of a second set of waves that accompany another such mode; at the same time the probability of finding the electron at any given distance from the nucleus must change because the intensity of the waves at that point has been altered. Thus there is a correspondence between the old particle picture and the newer viewpoint of wave mechanics.

275

Example. Find the de Broglie wavelength of an electron that has been accelerated through a potential difference $V = 1600$ V.

Solution. Since the electron's KE equals the work done on it during its acceleration, we have

$$\tfrac{1}{2}mv^2 = eV,$$
$$mv = \sqrt{2meV},$$
$$\lambda = h/mv = h/\sqrt{2meV}.$$

Taking $h = 6.6 \times 10^{-34}$ J-sec, $m = 9.1 \times 10^{-31}$ kg, $e = 1.6 \times 10^{-19}$ C, and $V = 1600$ V, we get

$$\lambda = 3 \times 10^{-11}\ \text{m} = 0.3\ \text{Å},$$

which is of the order of the separation of the planes in a crystal lattice.

11|8 THE UNCERTAINTY PRINCIPLE

We have seen that in wave mechanics all predictions are statistical, whereas in classical mechanics it was assumed that, given sufficient data, future events could be forecast with certainty, as in celestial mechanics. Of course, in atomic physics we do not have the "sufficient data." Attempts to simultaneously measure the position and the velocity of a particle always meet with frustration.

   Suppose, for example, that we decide to observe the position of an electron. To do so, we must let at least one photon of light be scattered by the particle. We know that an electron is too small to "see" in a microscope using visible light because no matter how large we may be able to make the magnification, we are limited by the resolving power, which in turn depends on the diffraction of the light. Since the resolving power is inversely proportional to the wavelength of the light, we may increase the practicable magnification by using radiation of shorter wavelengths, such as x-rays, but then the trouble is that one photon of such radiation strikes our electron with greater momentum and energy than does one photon of radiation of longer wavelength. We have seen in the Compton effect that when a photon is scattered by an electron, the latter recoils in some forward direction. The energy imparted to the electron increases with the energy of the incident photon. When we use radiation of shorter wavelength we improve our resolving power and thus may locate an object more accurately in *position*, but at the same time we increase the disturbance of the particle and make our knowledge of its *momentum* more uncertain (see Fig. 11–10).

If we use light of wavelength λ, the smallest distance that we can resolve is a separation of about one wavelength, or λ, so this becomes the uncertainty Δs with which we can determine the position of a particle. The momentum of one photon of our light is $hf/c = h/\lambda$ (Eq. 11–11) and since an appreciable but unknown fraction of this momentum is transferred to the particle being viewed, we may call this $\Delta\mu$, the uncertainty in our knowledge of the momentum of the particle. Note that

$$\Delta s \cdot \Delta\mu = \lambda \cdot h/\lambda = h.$$

The quantum theory thus informs us that it is meaningless to talk about a $\Delta s \cdot \Delta\mu$ less than h, so that in general we must have

$$\Delta s \cdot \Delta\mu > h. \qquad\qquad \textbf{11–17}$$

This is *Heisenberg's uncertainty principle*. The uncertainties referred to are inherent in the nature of light and matter, according to quantum mechanics; the usual errors due to imperfect equipment and measurement are additional to these.

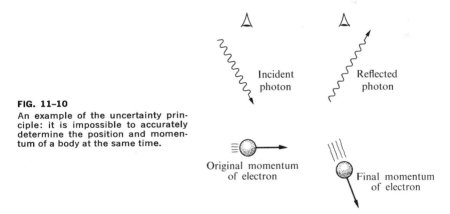

FIG. 11–10
An example of the uncertainty principle: it is impossible to accurately determine the position and momentum of a body at the same time.

We have seen that Planck's constant h has a finite value and that in the mks system its units are joule-seconds. The product of energy and time is called *action*, so that Planck's constant is known as *the quantum of action*. The quantum theory tells us that h is the smallest amount of action that has physical meaning.

It is the nature of things that, since Planck's constant has a finite value, infinitely detailed experience and sharp predictions are physically impossible. As a result of this inherent uncertainty, an event considered "certain" in classical physics becomes only "highly probable" in wave mechanics, and an event termed "impossible" in Newtonian theory is now classed as one having a "very low probability."

Example 1. The position of a 100-gm weight is measured with an uncertainty of ±0.1 mm. The weight is to be allowed to fall from "a state of rest." According to quantum mechanics it is meaningless to specify the initial velocity more specifically than to say that it may be Δv in any direction (up, down, or sideways), where $m\,\Delta v = \Delta\mu > h/\Delta s$. How large is Δv?

Solution. Here $m = 0.1$ kg, $\Delta s = 10^{-4}$ m, so that

$$\Delta v > \frac{h}{m\,\Delta s} = \frac{6.6 \times 10^{-34}\ \text{J-sec}}{0.1\ \text{kg} \times 10^{-4}\ \text{m}} = 6.6 \times 10^{-29}\ \text{m/sec}.$$

This is an uncertainty far below that which we could detect experimentally. In this case instrumental and personal errors would far outweigh the uncertainties of quantum mechanics.

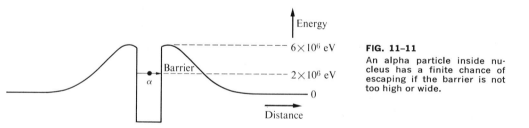

FIG. 11–11
An alpha particle inside nucleus has a finite chance of escaping if the barrier is not too high or wide.

Example 2. *The tunnel effect.* An alpha particle ($m = 6.4 \times 10^{-27}$ kg) may be constrained to stay inside of a certain nucleus by the attractive forces of other nuclear particles. The result may be pictured by visualizing the alpha particle as surrounded by a "hill" or barrier, as shown in Fig. 11–11. Suppose that the alpha particle has a KE = 2,000,000 eV but needs 4,000,000 eV additional energy to climb over the barrier and that the barrier has a width of 10^{-14} m (about the radius of a nucleus). So long as the alpha particle is known to be within the nucleus the uncertainty in its position cannot exceed a $\Delta s = 10^{-14}$ m. What, then, is the uncertainty in its velocity?

We have

$$\Delta v > \frac{h}{m \, \Delta s} = \frac{6.6 \times 10^{-34} \text{ J-sec}}{6.4 \times 10^{-27} \text{ kg} \times 10^{-14} \text{ m}} = 10^7 \text{ m/sec.}$$

Now, an alpha particle with a KE $= 2 \times 10^6$ eV has a velocity, v, of just this magnitude, so that

$$v = (10^7 \pm 10^7) \text{ m/sec.}$$

We see that its velocity *could* be 2×10^7 m/sec, in which case its KE would be 8×10^6 eV. (Remember that KE is proportional to v^2.) Since 8,000,000 eV exceeds the 6,000,000 eV needed to clear the barrier, we must conclude that the alpha particle has a chance of being found *outside* of the nucleus, just as though it had "leaked" out or found a "tunnel" through the barrier.

This is the quantum theory explanation of the process believed to occur when a sample of a radioactive element such as radium decays through the loss of alpha particles by its nuclei.

11|9 THE QUANTUM PRINCIPLE. FUNDAMENTAL LAW XIII

How shall we summarize all of the topics discussed in this chapter? The fact that each involves the same constant h implies that they are related through a common principle. The quantum theory embraces an area comparable with that of all classical physics, and it is for this reason that it cannot be summed up in a single sentence or equation.

Perhaps for the purpose of summarizing the facts discussed in this chapter the following statement may be taken as our *Fundamental Law XIII* relating to quantum phenomena.

Every particle, corpuscular or radiant, behaves as though it possesses both particle properties and wave properties, related as follows:

XIII. $\lambda = h/\mu$, $f = E/h$,   **11–18**

where h is a universal constant. It is further assumed that the particle properties are governed by the classical laws of Newton, Coulomb, etc., while the wave properties are those (such as diffraction and interference) described by Huygens' principle.

The particle model and the wave model, alone or together, do not give us the true and complete picture, but they may be of considerable help when it comes to describing or predicting the results of experiments in atomic physics. If one abandons pictorial models and turns to the mathematical and logical development of quantum mechanics one will find that it is a complete theory undisturbed by any wave-particle dilemma. It presents mathematical difficulties, but if these are overcome the results obtained are correct.

PROBLEMS

1. How many photons of light of wavelength equal to 2000 Å have the same total energy as

    a)   six photons of light of 6000 Å?

    b)   a photon of light of 3000 Å plus a photon of light of 6000 Å?

    c)   Calculate and tabulate the wavelengths in angstroms for which the energy per photon is, respectively, 1 eV, 2 eV, 3 eV, 4 eV, 5 eV, 10 eV, 100 eV, 10,000 eV, 1,000,000 eV.

2. When the solar spectrum was measured by instruments carried by rocket above the earth's atmosphere, it was found that λ_m for solar radiation is about 4700 Å (at the earth's surface the peak intensity is in the green region). Compute the temperature of the sun's surface, assuming that it radiates as would a blackbody.

3. a)   Find the ratio of the energy emitted by a blackbody at 3000°K to that emitted at 2000°K.

    b)   Find the ratio of the emission of a black steam radiator at 100°C to its emission at 27°C (about room temperature). °C = °K − 273°.

4. Light of wavelength λ strikes a photocell and photoelectrons of energy E are emitted.

    a)   What is the work function W of the metal in terms of λ, E, h, and c (the speed of light)?

    b)   Find, in terms of the same quantities, the energy E' of the photoelectrons emitted from the same metal by light whose wavelength is $\lambda/2$.

5. Show that Einstein's theory of the photoelectric effect can account for all of the experimental facts listed in Section 11–3.

6. In this problem you are not to assume the value of h but are to calculate it from the following data. Light from a sodium lamp falling on a photocell causes the emission of electrons for which the stopping potential is $V = 0.4$ V, while when violet light from a mercury arc is used, $V = 1.3$ V. Take $\lambda = 5.89 \times 10^{-7}$ m for sodium light and 4.05×10^{-7} m for the violet light used. (Refer to Experiment 18.)

    a) Compute h/e.

    b) Assume $e = 1.60 \times 10^{-19}$ C and find h.

    c) Compute the work function W of the metal used.

7. Compute λ in angstroms for the first line of the Lyman series and the first three lines of the Balmer series.

8. In an x-ray tube electrons are accelerated through a pd of V volts and then strike a metal target (see Fig. 11–12). Most of the electrons contribute their KE to the internal energy of the target, but some electrons that are stopped radiate photons of various wavelengths. What is the maximum energy (in electron-volts) that such photons can have? What effect is the converse of x-ray production?

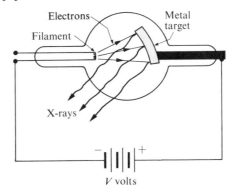

FIG. 11–12
X-ray tube. Electrons accelerated through a high potential difference strike a metal target.

9. Compute the shortest wavelength emitted by an x-ray tube operating at (a) 50,000 V, and (b) 250,000 V.

10. Explain why the recoil electrons in the Compton effect gain an insignificant amount of energy from photons of visible light.

11. Show that if Δt is the time taken by a particle in traversing the distance Δs and $E = \frac{1}{2}mv^2$, while $E + \Delta E = \frac{1}{2}m(v + \Delta v)^2$, then Eq. (11–17) becomes

$$\Delta E \cdot \Delta t > h.$$

Experiment 18 THE PHOTOELECTRIC EFFECT
Determination of Planck's Constant *h*

Object: To verify the quantum principle as applied to the photoelectric effect and to measure Planck's constant *h*.

Theory: The photoelectric effect was discovered by Hertz, and explained in terms of quantum theory by Einstein, whose theory was later verified experimentally by Millikan.

Suppose that a photon of energy hf gives its energy to an electron just inside the metal. This electron may use the energy to escape from the metal, but in so doing it must, so to speak, pay to get out. The work that must be done to escape is called the work function, W. The energy left after escape ($hf - W$) must be the kinetic energy of the electron; rather than measuring this KE directly, we shall let the electron turn its KE into potential energy as it moves through a potential drop of V volts. The gain in PE, or loss in KE, will be eV joules, so if V is just sufficient to stop the electrons,

$$hf - W = eV. \tag{1}$$

The stopping potential V can be measured directly.

We want to use Eq. (1) to determine h. We shall eliminate W by using successively two or more values of f and measuring the value of V for each. The best procedure is to plot V against f. Equation (1) predicts a straight line graph with a slope $\Delta V/\Delta f$ equal to h/e and an intercept on the V-axis at $-W/e$. Since e has been previously measured, h and W may be calculated.

Procedure

Step 1. As in Fig. 11–13, let light from a known monochromatic source strike the plate of your photocell. The electrons flow to a collector C and from it through the external circuit and back to the plate. The external circuit contains a galvanometer G to measure the electron current, and a variable voltage source (+2 to −2 V) with a voltmeter to measure V. Make the collector slightly negative and increase V until the galvanometer shows no current. Read V.

Step 2. Repeat 1 for two other known wavelengths. Suggested light sources are: sodium light ($\lambda = 5.89 \times 10^{-7}$ m), the green light filtered from a mercury arc ($\lambda = 5.46 \times 10^{-7}$ m), and the violet light from a mercury arc ($\lambda = 4.05 \times 10^{-7}$ m).

Step 3. Compute each frequency used from the relation $f = c/\lambda$, where c is the speed of light and λ the wavelength in meters. Plot V versus f and determine the slope and intercept of the line that best fits your experimental data.

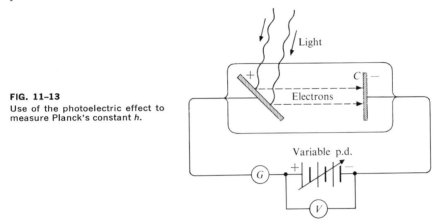

FIG. 11–13
Use of the photoelectric effect to measure Planck's constant h.

Step 4. Calculate h and compare your value with the accepted value of 6.63×10^{-34} J-sec. Can you justify your error?

Step 5. Calculate W in joules and in electron-volts. Note that 1 eV $= 1.6 \times 10^{-19}$ J.

HOFFMANN, B.,   *The Strange Story of the Quantum*, Dover, 1959.

GAMOW, G.,   *Thirty Years that Shook Physics*, Anchor, 1966.

References

12

12|1 INTRODUCTION Motorists are advised to remember that "two bodies cannot occupy the same place at the same time." What is implied in this statement is that the space, say at an intersection, occupied by car A at a given instant must be vacated by car A before car B can move into any portion of said space, provided that the cars are to remain intact. Of course solid objects may be jammed and telescoped together, or they may be compressed into smaller volumes, until two or more such objects occupy no more space than one did originally, but even in such cases we would still say that it is impossible for a piece of each object to pass through the same mathematical point at the same time.

Pauli's Exclusion Principle

Next consider liquids and gases. When a substance A is dissolved in a liquid B, the volume of the solution is less than the sum of the original volumes of A and B. Does this mean that some of A and some of B occupy the same space at the same time? They do in the macroscopic sense, but not from the microscopic point of view. We postulate that in a liquid the molecules are loosely packed and do not themselves occupy all of the available space, so that additional molecules may be fitted in between those of the liquid. In the case of a gas, the molecules behave as though they occupy an even smaller fraction of the available space than in the case of a liquid, for it is easy to compress a given amount of gas into a greatly reduced volume, or to add some more gas molecules to a given volume. On the molecular level we may ask whether or not two molecules can occupy the same space at the same time. If molecules behave like hard elastic spheres, the answer must be, "No." On the other hand, molecules may combine chemically, and when they do a new compound molecule may result from the partial merging of the original molecules. The molecules fit together as though they also contain within themselves some empty space.

Lord Rutherford (1871–1937) performed, around 1910, some crucial experiments which indicated that individual atoms behave as though they have a planetary structure. The mass of the atom seemed to be concentrated in a central positive nucleus (the atom's "sun"), around which negative electrons of much smaller mass (the atom's "planets") circulated in orbits. The diameter of the nucleus appeared to be about 10^{-15} to 10^{-14} m. Rutherford determined this by firing alpha particles (fast-moving nuclei of helium) through a foil that was so thin that one alpha particle passed close to only one atomic nucleus in the foil (see Fig. 12–1). He could estimate the distance of closest approach, and he found that if this distance exceeded a few times 10^{-15} m, the nuclei behaved as though they had not penetrated one another; for smaller distances penetration was evident.

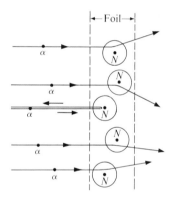

FIG. 12–1
Scattering of alpha particles by atomic nuclei (*N*) in a thin foil.

The diameter of an atom may be estimated from the average volume per atom of a solid. The apparent diameters of atoms are thus computed to be of the order of magnitude of 10^{-10} to 10^{-9} m or some hundred thousand times that of their nuclei. Evidently atoms themselves are largely empty space. Thus when atoms combine to form a molecule there is ample opportunity for the atoms to share some of their unoccupied space.

Example. The density of aluminum (Al) is 2700 kg/m^3 and its atomic weight is 27. Compute the size of the cube whose volume is the volume of a given mass of aluminum divided by the number of atoms in that mass.

Solution. Twenty-seven kg of Al (1 kg-mole) contain 6×10^{26} atoms. The mass of one Al atom is $27/6 \times 10^{26} = 4.5 \times 10^{-26}$ kg. Hence the number of atoms per m^3 is $(2700/4.5 \times 10^{-26})$ atoms/m^3. Thus the volume per atom is

$$\frac{4.5 \times 10^{-26} \, m^3}{2700 \text{ atoms}} = \frac{16.7 \times 10^{-30} \, m^3}{\text{atom}}.$$

For a cube of this volume the length of one edge would be $(16.7 \times 10^{-30})^{1/3}$, or about 2.5×10^{-10} m. If the atoms of Al were of cubical shape and completely occupied a given volume of the metal, 2.5×10^{-10} m (2.5 Å) would represent the length of one edge of such a cubic atom. If the atoms are not cubical or tightly packed, they may be somewhat different in size, but the fact that metals are nearly incompressible indicates that their atoms fill almost all of the available volume. The diameter of an atom is thus of the order of 10^{-10} m.

Do particles exist that truly behave like hard, incompressible spheres, two of which cannot occupy the same place at the same time? Perhaps the so-called "elementary particles" are of this type. The electron is classed as an elementary particle. Can two electrons occupy the same place at the same time? Considering Heisenberg's uncertainty principle, this last question is meaningless; the mass of the electron is so small that it is pointless to think of an electron as occupying an exact position in space at a given time. Nevertheless, there is definite evidence that there is a limit to how closely electrons may be crowded together, and the postulate that accounts for this evidence forms the fundamental principle of this chapter. The facts supporting this postulate will be discussed in Sections 12–4 through 12–7; to grasp the significance of these facts one must first understand what is meant by a "quantum state" and how the energy of an electron depends on the quantum state in which it is situated.

12|2 QUANTUM NUMBERS

We saw in the last chapter that the quantum number n, whose value could be $1, 2, 3, \ldots$, played an important role in both the Bohr theory and the wave-mechanical theory of the hydrogen atom. The normal state of lowest energy corresponded to $n = 1$, the next higher state corresponded to $n = 2$, etc., as shown in Fig. 11–6. It was explained that each value of n corresponds to a resonance of the waves associated with the electron in its orbit, just as there are various possible modes of vibration, or standing waves, in a string.

Further study of atomic spectra has revealed the necessity of introducing other quantum numbers to define quantum states. This is because atoms are three-dimensional systems and their allowed energy states correspond not so much to the possible standing waves in a one-dimensional string as they do to the possible resonance conditions in a small closed room. The allowable resonance states of such a room are determined by *three* sets of integers, rather than just one such set. So for atomic systems we must introduce at least three quantum numbers. Finally, when it was postulated that every electron has a spin in one of two opposite senses (see below) it became necessary to add a fourth quantum number, one limited to just two possible values. Thus it is found that the state of an electron in an atom may be described in terms of the following quantum numbers.

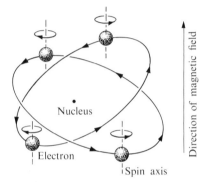

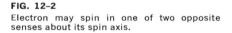

FIG. 12–2
Electron may spin in one of two opposite senses about its spin axis.

1. *The total quantum number, n*, whose value may be 1, 2, 3, ... The energy of a state depends primarily on the value of n.

2. *The orbital quantum number, l*, whose value may be 0, 1, 2, . . . , $(n - 1)$. Thus if $n = 3$, l may be 0, 1, or 2. The energy of a state depends to a lesser extent on l. If one wishes to adopt the planetary model of the atom one may consider l to be related to the eccentricity, or shape, of an electron orbit.

3. *The magnetic quantum number, m_l*, whose value may be l, $l - 1$, $l - 2$, . . . , $-l$. Thus if $l = 2$, m_l may be 2, 1, 0, -1, or -2. This quantum number is associated with the orientation (inclination) of the electron's orbit relative to the direction of the magnetic field of the nucleus. This magnetic field is explained by attributing a spin to the positively charged nucleus, somewhat analogous to the spin of the sun in our planetary system; such a rotating nucleus must have a magnetic field. Just as a magnetic compass needle has less energy when it swings around to its equilibrium position, in which it points north, than it does when it

points east or south, so the energy of an orbiting electron (which magnetically is equivalent to a small magnet) depends to some extent on the orientation of its orbit in the magnetic field within the atom.

4. *The spin quantum number, m_s,* whose value may be either $+\frac{1}{2}$ or $-\frac{1}{2}$. It is postulated that every electron has a spin of constant magnitude about an axis parallel to the magnetic field present and that the sense of the rotation about this axis may be one way (say clockwise to a certain viewer) or the opposite (counterclockwise), as in Fig. 12–2. The spin of the electron corresponds to that of a planet about its polar axis, with the difference that the planets are not alike and do not have the same spin, while electrons are alike and do have spins of identical magnitude and (in a given atom) the same or opposite direction.

12|3 QUANTUM STATES

A set of allowed values of the quantum numbers determines the so-called *quantum state* of a *bound electron* in an atom. The energy associated with a given quantum state varies with the values of n, l, m_l, and m_s, and it also depends on what other states are occupied in the same atom.

The *free electrons* in a metal are not restricted to motion around a particular nucleus, yet these electrons are not completely free because they cannot normally escape from the metal. These electrons are confined to the interior of the metal and in this respect they resemble molecules in a box. Due to this restriction, only certain modes of vibration are possible for the waves associated with the free electrons, just as there are only certain resonant acoustical frequencies associated with a small enclosed space. As a result, the energy of each electron is restricted to a discrete set of values which, on an energy-level diagram, lie one above another like the rungs of a ladder. Thus we may speak of an electron, bound or free, as being in a certain *state*.

12|4 EVIDENCE THAT ELECTRONS CANNOT BE CROWDED INTO THE QUANTUM STATE OF LOWEST ENERGY

We saw in Chapter 5 that our universe seems to be "running down." If a group of fast (hot) molecules is mixed with a group of slow (cold) ones, the more energetic molecules share, through collisions, their excess energy with the less energetic ones until the two groups are indistinguishable as far as their energy distribution is concerned. We did not go into the study of what the energy distribution is for gas molecules in thermal equilibrium, but found only that the mean KE of translation per molecule is $\frac{3}{2}kT$. Actually some molecules have more and some less than this average energy; a large percentage have practically no energy, while a few have many times the average amount. For gas molecules there seems to be no restriction other than temperature as to how many of them may be in the quantum state of lowest energy, but for electrons the situation appears to be quite different.

We shall consider three important lines of evidence which suggest that electrons cannot be crowded into the state of lowest energy regardless of how low the temperature may be. These three topics are: (1) the heat capacities of metals, (2) atomic diameters, and (3) the periodic table of the elements.

12|5 THE HEAT CAPACITIES OF METALS

We picture a metal as a latticelike arrangement of atoms, or ions (atoms that have lost one or more of their electrons), between which circulate free electrons. At room temperature the atoms are considered to be oscillating about their respective lattice points. To raise the temperature of a metal, one must add energy, usually by heating it, so as to increase the thermal agitation of atoms about the lattice points. One would expect that if the free electrons behave like gas molecules, their KE would also have to be increased as the temperature is raised, but it is found experimentally that this is not the case!

The average KE of translation of a gas molecule is $\frac{3}{2}kT$, regardless of its mass, so that when T is raised by one degree the mean KE of translation per molecule is raised by $\frac{3}{2}k$. The oscillating atoms or ions of a crystal lattice possess this same mean KE, and in addition an equal average potential energy, so that altogether each must be given the energy $3k$ to raise its temperature one degree. Since $3k$ times the number of atoms per kg is about the measured value of the specific heat of most metals at ordinary temperatures, one must conclude that the free electrons do not appear to take any of the energy that must be given to a metal when its temperature is raised. To see how this might be possible, consider the following analogy.

Ski Club A sends a group of men to a resort where there is a large hotel. This hotel has available one dormitory room with a dozen bunks that rent for $1.00 a night, another large room with cots at $2.00 a night, and better rooms with beds at $3.00 per person, $4.00 per person, and up. There are twelve men in the group and they have $16.00 between them to spend per night. They might decide to put eight men in bunks and four on cots, or to put eleven men in bunks and then give their leader a $5.00 bedroom. (Can you think of other possible arrangements that total $16.00?) Now suppose that unexpectedly the group meets a wealthy friend of the club and that this person offers each of the twelve members in turn one dollar to improve the respective member's sleeping accommodations. Each person gratefully accepts the dollar and moves up to the next higher-priced accommodation. The friend thus contributes $12.00 in all.

Ski Club B sends twelve men to another resort where the hotel has a similar bunkroom and cot-room plus *one* two-bed room at $3.00 per person per night, *one* at $4.00, and so on up. This group of skiers is peculiar in that no two of them will share a room. The cheapest arrangement possible is to put one man in the bunkroom, one in the cot-room, one in the $3.00 bedroom, one in the $4.00 bedroom etc., with the leader taking a bed in the $12.00 room. This totals $78.00 in all, so Club B allots this amount to its group per night. This group also meets a gentleman who approaches each of them privately and makes the same offer of one dollar to improve the member's accommodations. The man in the bunkroom refuses to move up to the cot-room where he knows another member is lodged. When the would-be benefactor speaks to the man in, say, the $4.00 room, this man explains that he would prefer not to move up into the $5.00 room as he would then have to share it with the member assigned to it. The cheapest unoccupied room costs $13.00 and the benefactor is not willing to contribute the $9.00 needed to move the skier in the $4.00 room into this expensive room. So the benefactor only succeeds in helping the leader move from his $12.00 room to the $13.00 room. Unfortunately, the men do not all get together in time to realize that with $12.00 extra they could have moved simultaneously, each going up one notch on the room scale. Therefore, in this case the group actually gains little from the benefactor.

In this analogy, dollars per bed correspond to the energy associated with a particle in a given state. The members of Club A, who will gladly share a bedroom, correspond to the atoms of a metal and the members of Club B, who must sleep in separate rooms, correspond to the free electrons. The contribution of the wealthy friend is analogous to giving the metal sufficient energy to raise its temperature, say by one degree. In other

291

words, suppose that only one electron can occupy each possible energy state; then the distribution leading to the lowest total energy (such as one would expect near absolute zero) is that where the states of lowest energy are all solidly filled, each with one electron. Enough of these states from zero energy up to, say, E_m must be occupied to accommodate all of the electrons, as shown in Fig. 12–3(a). States of energy above E_m will be empty.

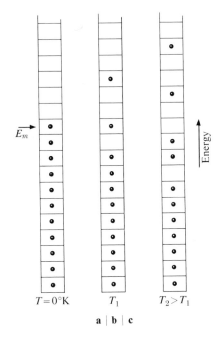

FIG. 12-3
Distribution of electrons among possible energy states at three different temperatures.

Picture a metal near absolute zero being given enough energy to raise its temperature one degree. Very little of this energy will be transferred to the electrons in the metal, for the average thermal energy of the atoms ($3kT$) is not enough to disturb those electrons which are not near the top of the filled states. Such electrons would have to jump all the way up to the unfilled states above E_m. However, a little energy might be absorbed by electrons in those states just below the highest one filled, raising them to nearby unfilled states, as in Fig. 12–3(b) and (c). The fraction of the free electrons moving to higher energy states is small at ordinary temperatures, and thus the free electrons contribute little to the heat capacity.

12|6 ATOMIC DIAMETERS

In the Example of Section 12–1, the diameter of an Al atom was calculated from the density of the metal to be no greater than 2.5×10^{-10} m, or 2.5 Å. By the same method one finds the maximum possible diameter of a uranium (U) atom to be 2.8 Å. There are more exact methods for computing the size of atoms, but they all indicate that atoms of large atomic weight are of approximately the same size as those of small atomic weight. If one compares elements of similar chemical properties, such as the alkalis, one finds a steady *increase* in diameter with atomic weight, as follows: 1.56 Å for lithium (Li), 1.86 Å for sodium (Na), 2.23 Å for potassium (K), 2.36 Å for rubidium (Rb), and 2.55 Å for cesium (Cs). (These values are based on the best methods of computation.) This increase in diameter is not what one would expect if all of the orbital electrons of an atom are in the quantum state of lowest energy.

From his work on the scattering on alpha particles by nuclei, Rutherford deduced among other things that the positive charge on the nucleus of an atom increases steadily from $+e$ for hydrogen to $+92e$ for uranium. Thus the nucleus of a heavy atom must be surrounded by a more intense electric field than that around the nucleus of a light atom. A stronger field should pull an electron that is orbiting in the quantum state of lowest energy (the one for which $n = 1$) in closer to the nucleus. Hence, if all of the orbital electrons of an atom could exist in the quantum state of lowest energy, the atoms of large atomic weight and nuclear charge, such as those of uranium, should be the smallest in diameter. Since this is not the case, we must conclude that all of the electrons in an atom cannot be in the lowest quantum state, but must occupy states of different energies. Again we see evidence of the existence of a principle that forbids the crowding of electrons into the same quantum state.

12|7 THE PERIODIC TABLE OF THE ELEMENTS

We have noted that certain groups of elements of widely different atomic weights, such as the alkalis, have similar chemical properties. The alkalis form one such group, the alkaline earths (Be, Mg, Ca, Sr, Ba, Ra) another, and the halogens (F, Cl, Br, I) still another. This recurrence of similar chemical properties as one goes from the lighter to the heavier elements was the subject of much speculation. Finally, in 1869, the great Russian chemist Dmitri Mendeléeff (1834–1907) hit upon a fruitful scheme for

TABLE 12-1
Periodic Classification of the Elements (Mendeléeff, 1872)

Group →		I	II	III	IV	V	VI	VII	VIII
Series	1	H							
	2	Li	Be	B	C	N	O	F	
	3	Na	Mg	Al	Si	P	S	Cl	
	4	K	Ca	—	Ti	V	Cr	Mn	Fe, Co, Ni, Cu
	5	[Cu]	Zn	—	—	As	Se	Br	
	6	Rb	Sr	Yt?	Zr	Nb	Mo	—	Ru, Rh, Pd, Ag
	7	[Ag]	Cd	In	Sn	Sb	Te	I	
	8	Cs	Ba	Di?	Ce?	—	—	—	
	9	—	—	—	—	—	—	—	
	10	—	—	Er?	La?	Ta	W	—	Os, Ir, Pt, Au
	11	[Au]	Hg	Tl	Pb	Bi	—	—	
	12	—	—	—	Th	—	U		

classifying and arranging the then known elements. His method consisted of writing down the symbols of the elements in such an order that the atomic weights of the respective elements steadily increased throughout the series; as he did this, Mendeléeff placed elements with similar chemical properties in the same vertical column. He then found that if gaps were left for presumably "missing" elements, a regular periodicity occurred. Each even series of his table, which is reproduced in Table 12-1, commences with an alkali element, followed by an alkaline earth; the third elements in each row resemble one another, as do the ones in the fourth column, the fifth column, etc.

The chemical properties of an element are determined by the ability of its atoms to combine with those of other elements. A halogen atom has the capacity to *combine with one atom of hydrogen* to form molecules such as HCl, HBr, etc., while the elements of the sixth column, or group, con-

TABLE 12–2
Valences of Some Common Elements

Element	Symbol	Valence
Hydrogen	H	$+1$
Lithium	Li	$+1$
Sodium	Na	$+1$
Potassium	K	$+1$
Beryllium	Be	$+2$
Magnesium	Mg	$+2$
Calcium	Ca	$+2$
Zinc	Zn	$+2$
Boron	B	$+3$
Aluminum	Al	$+3$
Carbon	C	$(+4), -4$
Silicon	Si	$(+4), -4$
Nitrogen	N	$(+5), -3$
Oxygen	O	-2
Sulphur	S	$(+6), -2$
Chlorine	Cl	$(+7), -1$
Bromine	Br	$(+7), -1$
Copper	Cu	$+1, +2$

sist of atoms that may combine with two atoms of hydrogen to form molecules such as H_2O, H_2S, etc. An alkali atom may *replace one hydrogen atom* in a molecule, thus forming molecules like LiCl, NaCl, KBr, etc. An alkaline earth atom may replace two hydrogen atoms in a molecule, and so on. The term *valence* has been introduced to summarize what has just been said. Valence measures the capacity of atoms to react with one another. *Positive valence* is the number of hydrogen atoms which an atom of the element may replace, and *negative valence* is the number of hydrogen atoms with which an atom of the element may combine. Some elements have more than one valence, such as a positive valence of v and a negative valence of $(8 - v)$. The valences of some common elements are listed in Table 12–2. We see that as one passes from the first group to the seventh in Mendeléeff's table, the positive valence increases from 1 on up, while the negative valence becomes less negative and reaches -1 for the seventh group, the halogens.

Mendeléeff's table was an empirical accomplishment; it summarized many facts, and this alone made it important. It serves as a table to which one may refer when one wishes to recall the general chemical properties of any element. However, the importance of Mendeléeff's achieve-

TABLE 12-3
Modern Version of the Periodic Table

Outer electrons are in the	I	II	III	IV	V	VI	VII	VIII	O	Electrons per shell
First or K-shell	1 H 1.0080								2 He 4.003	2
Second or L-shell	3 Li 6.940	4 Be 9.013	5 B 10.82	6 C 12.011	7 N 14.008	8 O 16.000	9 F 19.00		10 Ne 20.183	2, 8
Third or M-shell	11 Na 22.991	12 Mg 24.32	13 Al 26.98	14 Si 28.09	15 P 30.975	16 S 32.066	17 Cl 35.457		18 Ar 39.944	2, 8, 8
Fourth or N-shell	19 K 39.100	20 Ca 40.08	21 Sc 44.96	22 Ti 47.90	23 V 50.95	24 Cr 52.01	25 Mn 54.94	26 Fe 55.85 / 27 Co 58.94 / 28 Ni 58.71		2, 8, 18, 8
	29 Cu 63.54	30 Zn 65.38	31 Ga 69.72	32 Ge 72.60	33 As 74.91	34 Se 78.96	35 Br 79.916		36 Kr 83.80	
Fifth or O-shell	37 Rb 85.48	38 Sr 87.63	39 Y 88.92	40 Zr 91.22	41 Nb 92.91	42 Mo 95.95	43 Tc (98)	44 Ru 101.10 / 45 Rh 102.91 / 46 Pd 106.4		2, 8, 18, 18, 8
	47 Ag 107.880	48 Cd 112.41	49 In 114.82	50 Sn 118.70	51 Sb 121.76	52 Te 127.61	53 I 126.91		54 Xe 131.30	
Sixth or P-shell	55 Cs 132.91	56 Ba 137.36	57-71 La series*	72 Hf 178.50	73 Ta 180.95	74 W 183.86	75 Re 186.22	76 Os 190.2 / 77 Ir 192.2 / 78 Pt 195.09		2, 8, 18, 32, 18, 8
	79 Au 197.0	80 Hg 200.61	81 Tl 204.39	82 Pb 207.21	83 Bi 209.00	84 Po (209)	85 At (210)		86 Em (222)	
Seventh or Q-shell	87 Fr (223)	88 Ra 226.05	89 — Ac series**							

*Lanthanide series:	57 La 138.92	58 Ce 140.13	59 Pr 140.92	60 Nd 144.27	61 Pm (145)	62 Sm 150.35	63 Eu 152.0	64 Gd 157.26	65 Tb 158.93	66 Dy 162.51	67 Ho 164.94	68 Er 167.27	69 Tm 168.94	70 Yb 173.04	71 Lu 174.99	2, 8, 18, 32, 9, 2
**Actinide series:	89 Ac (227)	90 Th 232.05	91 Pa (231)	92 U 238.07	93 Np (237)	94 Pu (242)	95 Am (243)	96 Cm (247)	97 Bk (247)	98 Cf (251)	99 Es (254)	100 Fm (253)	101 Md (256)	102 No (253)	103 Lw (253)	2, 8, 18, 32?, 9?, 2

ment is enhanced by the fact that it (1) made successful predictions, and (2) stimulated work on a theory to "explain" the periodicity of the table. Recall that in a similar manner Kepler's empirical laws summarized planetary motion and led to Newtonian theory; the Bode-Titus law for the radii of planetary orbits helped lead to the discovery of new planets; the laws of blackbody radiation led to Planck's quantum theory; and Balmer's formula for the wavelengths of the lines in the hydrogen spectrum resulted in Bohr's important work.

Mendeléeff wisely left spaces in his table for "missing" elements. He predicted the chemical and physical properties of each missing element, which helped lead to the eventual discovery of these elements. Now all of the gaps in the table have been filled, either by elements found to occur naturally in the earth, or by elements made artificially through nuclear transmutation, the latter including over ten elements beyond uranium (the transuranic elements). The original table has been greatly modified, by the addition of Group 0, which contains the inert gases (He, Ne, Ar, Kr, Xe, Rn) whose valence is zero. These inert elements were not known in Mendeléeff's time. Another change is the addition of the lanthanide rare earth elements between Ba and Hf, and a similar series following radium. It has also been found necessary in certain cases to disregard the rule that elements are to be arranged in order of increasing atomic weight; thus argon (Ar), an inert gas, obviously belongs under helium (He) and neon (Ne), while potassium (K) belongs under lithium (Li) and sodium (Na), which puts Ar ahead of K in spite of the fact that Ar has the greater atomic weight of the two.

A modern version of the periodic table is shown in Table 12-3. The elements are numbered in order and this position number is called the *atomic number Z*. One of the most important features of this modern table is that the periods are not of equal length. A *period* may be defined as any consecutive group of elements starting with hydrogen or an alkali and ending with an inert gas. The successive periods contain 2, 8, 8, 18, 18, and 32 elements, respectively. Note that $2 = 2(1)^2$, $8 = 2(2)^2$, $18 = 2(3)^2$, $32 = 2(4)^2$.

We now come to the important question of what the significance of all this is. Why does this periodicity of the elements occur, and why do we encounter the "magic numbers" 2, 8, 18, and 32? Is there some fundamental principle? Physicists are generally inclined to believe that in such a situation a fundamental law of nature must be involved. In this case the principle was first stated by the Austrian-born physicist Wolfgang Pauli in 1925. Pauli's postulate "explains" the periodic table and also the energy distribution of the free electrons in a metal and the variations in atomic diameters that we discussed earlier.

12|8 PAULI'S EXCLUSION PRINCIPLE. FUNDAMENTAL LAW XIV

Pauli's important postulate was simply the following:

No two electrons in a given system can be in the same quantum state.

The word *system* is here taken to mean either an atom or a certain volume of a metal. Since the quantum state of an electron in an atom is determined by the values of the four quantum numbers, n, l, m_l, and m_s, Pauli's principle asserts that *no two electrons in an atom can have the same set of quantum numbers.*

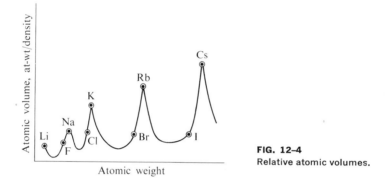

FIG. 12–4
Relative atomic volumes.

Pauli's postulate immediately predicts that the free electrons in a metal must be distributed among the quantum states of various energies as shown in Fig. 12–3. If one one electron can occupy each state, the behavior of the free electrons will be analogous to that of the skiers in Club *B* who refused to share the same room. It was explained that this sort of behavior would result in the free electrons making almost no contribution to the heat capacity of a metal. In other words, if a metal is given heat to raise its temperature, only a slight amount of this heat goes into increasing the energy of the free electrons and almost all of it goes into raising the internal energy of the atoms or ions of the metal.

The nucleus of any of the heavier atoms is surrounded by many electrons. Pauli's principle states that these electrons cannot all be in the state of lowest energy, i.e., the state for which the total quantum number n equals one. If the possible energy states are filled, each with one electron, in order of increasing energy, just as they are in the case of a metal, then electrons must be located in states for which $n = 2$, 3, or more. Higher values of n correspond to greater distances from the nucleus.

On the other hand, the stronger electric field of a heavy nucleus pulls all of the orbiting electrons in closer. The result of these two phenomena, which have opposing effects in regard to the size of an atom, is that light and heavy atoms of elements in the same group differ little in size. The diameter of an atom of a given element depends more on the group in the periodic table to which the element belongs than it does on the period in which the element is found (see Fig. 12–4).

TABLE 12–4
Possible Different Quantum States for $n = 1$ and $n = 2$

n	l	m_l	m_s	Number of possibilities
1	0	0	$\frac{1}{2}$	
1	0	0	$-\frac{1}{2}$	2
2	0	0	$\frac{1}{2}$	
2	0	0	$-\frac{1}{2}$	
2	1	1	$\frac{1}{2}$	
2	1	1	$-\frac{1}{2}$	
2	1	0	$\frac{1}{2}$	8
2	1	0	$-\frac{1}{2}$	
2	1	-1	$\frac{1}{2}$	
2	1	-1	$-\frac{1}{2}$	

Finally we come to the explanation of the "magic numbers" 2, 8, 18, and 32. In order to count the number of possible atomic states associated with $n = 1$, $n = 2$, $n = 3$, etc., it is best to list all of the different combinations of the four quantum numbers, as in Table 12–4. (See also problem 4 at end of chapter.) The "magic numbers" are the numbers of possible states for $n = 1, 2, 3, 4, \ldots$, respectively.

12|9 A REVIEW OF THE CHEMICAL ELEMENTS

For the first two elements in the periodic table, hydrogen (H) and helium (He), the electrons may be in $n = 1$ states (Fig. 12–5). In the case of the He atom, which normally has two electrons, the $n = 1$ states are filled. When we come to the lithium (Li) atom, which has a third electron, we see that this electron must be located in an $n = 2$ state, a state of considerably higher energy. A state of higher energy is one of lower stability, therefore a Li atom should be more prone to part with this

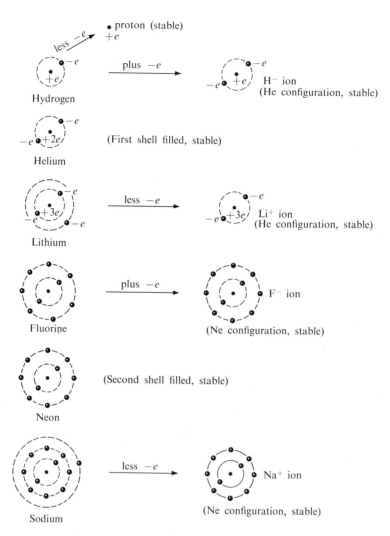

FIG. 12-5. Representative electron structures.

electron and become a Li$^+$ ion (Fig. 12–5). The result of this is that Li can easily enter into chemical combination with elements whose atoms seek an extra electron, whereas He cannot. Since with He the $n = 1$ states are filled and closed to further electrons, we think of He as marking the end of the first period of the periodic table of the elements. The electrons in the $n = 1$ state are said to form a closed "shell," but the term "shell" should not be taken too literally.

With the tenth element, neon (Ne), the $n = 2$, or second, shell of electrons is filled. The eight electrons in this shell are divided as follows: two in $l = 0$ states and six in $l = 1$ states (see Table 12–4). From the inert chemical property of Ne we deduce that this arrangement is also a particularly stable one.

Fluorine (F) is the element preceding Ne and its atom needs one more electron to complete its second shell, hence F is an example of the type of element with which Li will combine. The sodium (Na) atom, on the other hand, normally possesses one more electron than does the Ne atom, this last electron being in an $n = 3$ state outside of the completed second shell. The sodium atom tends to give up this electron and become a positive ion, Na^+ (Fig. 12–5). Thus Na and Li have similar chemical properties.

With the eighteenth element, argon (Ar), the $n = 3$, or third, shell of electrons also is found to contain two electrons in $l = 0$ states and six in $l = 1$ states. Since the electrons in the outermost shell are the ones that are most readily available for chemical combination, we may reason that because of the similar 2 and 6 arrangements of electrons in their outermost shells, Ar and Ne should have similar chemical properties, which they do. Argon is also an inert gas. Thus the third period of the periodic table also contains 8 (and not 18) elements.

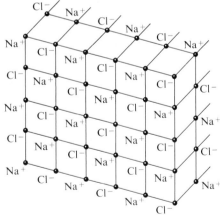

FIG. 12–6
Cubic lattice structure of sodium chloride crystal.

The atoms of chlorine (Cl), the element preceding Ar, need one more electron to complete the third shell. A chlorine atom may accept such an electron from, say, a sodium atom, and become a negative Cl^- ion. Common table salt is composed of sodium chloride (NaCl). This is a crystalline solid at ordinary temperatures and its crystal structure is represented by a cubic lattice in which sodium and chlorine ions alternate, as shown in Fig. 12–6. While some energy is required to pull an electron away from a sodium atom against the attraction of the positive nucleus, this energy is small because (1) the electron is in an outer orbit and (2) the nucleus is to a large extent shielded by the remaining electrons. This expenditure of energy is more than compensated for by the energy released as the electron is pulled in by the chlorine atom and as the resulting Na^+ and Cl^- ions are drawn closer together by their electrostatic attraction (see Coulomb's law). Thus the process

$$Na + Cl \rightarrow Na^+ + e^- + Cl \rightarrow Na^+ + Cl^-$$

can proceed with the release of energy. In the crystalline state shown in Fig. 12–6, we find that a large number of oppositely charged ions have been pulled together to form a "supermolecule." Thus sodium chloride is a very stable substance. When dissolved in water the electrostatic attractions between the Na^+ and Cl^- ions are weakened and the two kinds of ions may break away from one another and move separately through the water. These ions may serve as conductors of electric current, so that salt solutions are good conductors of electricity. If NaCl dissociated into neutral atoms, a salt solution would not be a good conductor.

Let us return to the periodic table of the elements. It is found that in the elements beyond Ar the additional electrons are added according to two principles, namely (1) Pauli's principle of one electron per allowed state and (2) a state of lower energy is normally filled before one of higher energy. Because of (2) it turns out that states of higher n- and lower l-values may thus be favored over those of lower n- and higher l-values. With the 36th element, krypton (Kr), also an inert gas, the $n = 3$ shell is filled and the $n = 4$ shell contains two electrons in $l = 0$ states and six in $l = 1$ states. Thus the period from potassium (K), the element after Ar, to Kr contains 18 elements and is represented by the addition of 10 electrons in $n = 3$, $l = 2$ states, 2 electrons in $n = 4$, $l = 0$ states, and 6 electrons in $n = 4$, $l = 1$ states. The next period also contains 18 elements and is accounted for by the addition of 10 electrons in $n = 4$, $l = 2$ states, 2 in $n = 5$, $l = 0$ states, and 6 in $n = 5$, $l = 1$ states.

The long period of 32 elements following xenon (Xe) and ending with radon (Rn) is represented by the addition of 14 electrons in $n = 4$, $l = 3$ states, 10 in $n = 5$, $l = 2$ states, 2 in $n = 6$, $l = 0$ states, and 6 in $n = 6$, $l = 1$ states. A similar period of 32 elements should end with element number 118, but such an element is far beyond the range of stable elements and even beyond those unstable ones that have been produced artificially.

We see that with the aid of Pauli's principle a complete explanation of the periodic table is possible.

The *electrostatic* or *ionic* type of *bond* that accounts for the stability of NaCl is not the only one that holds atoms together in molecules. Even more common is the so-called *covalent bond*, which involves the sharing of electrons in pairs by neighboring atoms in a molecule. With the inclusion of the shared electrons both atoms of the pair attain the stable structure of eight (two for hydrogen) electrons in the outermost shell. See Fig. 12–7.

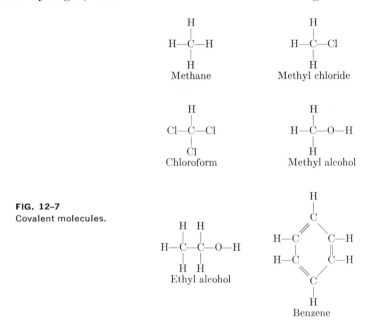

FIG. 12–7
Covalent molecules.

A further discussion of bonding is left to chemistry and lengthier physics texts. However, in summarizing we may say that chemical bonding, which is the heart of chemistry, may be explained in terms of fundamental physical laws, particularly in terms of Pauli's exclusion principle, the quantum principle, and Coulomb's and Ampère's laws. Of these principles that of Pauli plays the most significant role.

PROBLEMS

1. Take the diameter of the hydrogen molecule to be 1.4×10^{-10} m. A kg-mole (6×10^{26} molecules) of hydrogen (molecular weight $= 2$) at 0°C and 1 atm occupies 22.4 m³. What fraction of the occupied space is taken up by the molecules themselves?

2. a) Prove that if two molecules collide as hard elastic spheres of radii R, the center of one molecule is excluded by the other from a volume equal to eight times the volume of either molecule.

b) Oxygen gas behaves in a way that indicates that this excluded volume is 0.032 m^3 for 3×10^{26} *pairs* of molecules. Find the radius of the oxygen molecule.

3. Compute the volume and approximate size of a uranium atom from the density $(18.6 \times 10^3$ kg/m$^3)$ and atomic weight (238) of uranium.

4. List the values of l, m_l, and m_s for all the quantum states of an atom for which $n = 3$ and $n = 4$.

5. The 14 elements following La are called the "rare earths"; they are similar chemically because they differ only in the number of inner electrons that each possesses. What are the quantum numbers for the states filled by these inner electrons? Why are there 14 of these rare earth elements?

6. Why does copper, an element in Group I of the periodic table, differ chemically from the alkalis, Li, Na, K, etc.?

7. Indicate what ions are formed when each of the following salts is dissociated:

a) KI,   b) CuCl,   c) CuCl$_2$,   d) MgO,

e) AlCl$_3$,   f) Al$_2$O$_3$.

8. Common salt is formed in the reaction

HCl + NaOH → NaCl + H$_2$O.

What is the physical explanation of this reaction? In a salt solution, which ionizes more readily, the salt or the water?

Experiment 19 MONOMOLECULAR LAYERS
Determination of the Approximate Size of a Molecule

Object: To form a film one molecule thick and then to determine its thickness.

Theory: Atomic physics provides much indirect evidence that atoms and molecules have dimensions of around 10 Å units, or 10^{-9} m. In this experiment we shall measure the length of a molecule directly.

Because of their polar properties, the class of organic compounds known as fatty acids lend themselves quite effectively to monomolecular layer studies. As early as 1917, Langmuir pointed out that such molecules must line up vertically in a monomolecular layer, that is, with their longest dimension perpendicular to the surface of the layer, which, if unconfined, will be one molecule thick. Then the length of such a molecule may be determined if the thickness and area of the layer are known. In this experiment we shall try to form a monomolecular layer with a circular periphery, then we shall compute its volume V and area A. The thickness t will be given by

$$t = V/A. \tag{1}$$

Procedure

Step 1. Clean a large tray-shaped vessel thoroughly and fill it with distilled water. Cover the surface with powder or chalk dust.

Step 2. Calibrate a medicine dropper, that is, measure the volume of one of its drops by measuring the volume of 20 or more drops.

Step 3. Compute roughly the volume of fatty acid one drop should contain in order that it will form a circular area one molecule thick and around 10 cm in diameter. (Assume the thickness to be approximately 2×10^{-9} m.) Now compute how much the fatty acid should be diluted. Prepare a solution of this strength. Oleic or stearic acid diluted with benzene or alcohol (which evaporates quickly when spread on water) should work well. Let one drop of the dilute solution fall on and spread across your powdered water surface, then measure its diameter. Compute the thickness t of the monomolecular layer.

13|1 INTRODUCTION The assumption that matter is indestructible is a postulate of long standing. When processes that might be viewed as exceptions to this principle have been encountered, scientists have "saved" the principle by defining other forms of matter. (In a similar manner, the conservation of energy principle has been maintained by inventing different forms of energy.) The evolution in our thinking about matter has led to the realization that what is conserved in regard to matter is not its volume or its mass, but its constituent particles. Yet with the discovery of pair production and annihilation and our recent knowledge about the reactions between elementary particles, even the concept of particle conservation has

The Conservation of Matter Principle

required modification. Let us follow step by step this evolutionary development of the conservation of matter principle.

13|2 THE CONSERVATION OF MATTER IN PHYSICAL PROCESSES

If a child is playing with marbles and one gets lost, he soon learns to assume that the missing marble still exists somewhere, even though he cannot find it. Time and again lost objects do eventually turn up. The idea that objects such as marbles really vanish into, or materialize from, nowhere is regarded as superstitious.

Should an object like a marble break, we assume that the sum of the parts equals the original whole. This postulate is often substantiated experimentally. We believe that the fragments contain all of the molecules that the object possessed when whole.

When water boils it disappears from sight, but we postulate that the water molecules still exist in the form of an invisible gas which is called water vapor. This postulate is supported by the further fact that if water is boiled in an enclosed container, so that the water vapor cannot escape out into the atmosphere, then the visible water may be recovered by cooling and condensing the vapor (Fig. 13–1). We have seen how pressure is attributed to molecular impacts; water vapor also exerts a pressure on the walls of its container, just as air and other gases do, and so the H_2O molecules must still exist in the vapor.

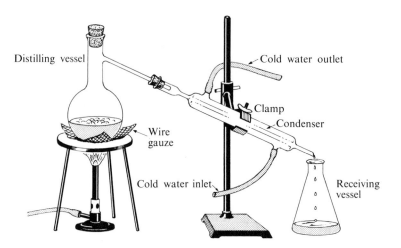

FIG. 13–1. Boiling and condensation of a liquid. Any solid dissolved in a distilling liquid will eventually recrystallize out.

When sugar and salt are dissolved in water they too seem to disappear; here again we postulate that their molecules are not destroyed but rather that they exist in a dissolved state, as witnessed by the change in taste. This assumption also explains why dissolved salt may be recovered in the original solid state by evaporating or boiling away the water.

In processes such as the above, in which molecules are neither gained nor lost, mass also appears to be conserved. Thus when 1 kg of water is boiled away, 1 kg of water vapor is produced. Actually the mass of the water vapor should, according to the relativity principle, be a *little* greater than that of the water from which it came. This follows from Einstein's relationship $E = mc^2$, according to which one must associate the mass E/c^2 with the energy E. Let us write

1 kg-mole of water (liquid) $+ Q =$ 1 kg-mole of water vapor, **13–1**

where Q is the heat of vaporization per kg-mole, i.e., the heat energy that must be transferred to one kg-mole of water to vaporize it. The value of Q varies somewhat with temperature, but at 100°C it is close to 4×10^7 J/kg-mole. One kg-mole of water is 18 kg of water. The mass equivalence of Q is

$$m = \frac{Q}{c^2} = \frac{4 \times 10^7 \text{ J}}{9 \times 10^{16} \text{ m}^2/\text{sec}^2}$$

$$= 4.4 \times 10^{-10} \frac{\text{kg-m}^2/\text{sec}^2}{\text{m}^2/\text{sec}^2}$$

$$= 4.4 \times 10^{-10} \text{ kg}.$$

This is obviously far too small a fraction of the mass to be observable.

13|3 THE CONSERVATION OF MATTER IN CHEMICAL PROCESSES

In chemical reactions atoms and electrons are rearranged, but not destroyed. For example, when sodium hydroxide, a base, is neutralized with hydrochloric acid, the reaction is as follows:

$$\text{NaOH} + \text{HCl} \rightarrow \text{NaCl} + \text{H}_2\text{O} + Q. \qquad\qquad \textbf{13–2}$$

We conclude that in chemical reactions the number of *molecules* of a given kind does change, but not the number of *atoms*. So we must modify our previous conservation of molecules principle and convert it into a conservation of atoms principle.

The Q in Eq. (13–2) represents the *heat of reaction*, or the chemical energy released as a result of the rearrangement of the valence electrons so as to form new and stronger bonds. When Q is on the right-hand side of the equation the reaction is termed *exothermal* if Q is positive, for in this case chemical energy is released; if Q is negative, the reaction is said to be *endothermal* and energy must be supplied from outside to make the process proceed as indicated.

Our modern atomistic view of chemical reactions dates back to the beginning of the 19th century. Before that time it was difficult for people to believe that the conservation of matter principle applied to chemical processes. People saw that when a candle burned its substance and mass obviously decreased, while when a metal was allowed to rust its mass increased. Then the discovery of oxygen threw new light on the process of combustion. In 1804 Dalton, an English schoolteacher, postulated that every elementary substance is composed of indestructible atoms and that the atoms of a given element are all alike and hence possess the same mass. The atomic weights, or the relative atomic masses, of the known elements were determined. The atomic weight of carbon is 12 and that of oxygen is 16. When a candle burns we say that its carbon atoms combine with the oxygen molecules (two atoms each) in the air to form carbon dioxide, an invisible gas. The reaction is

$$C + O_2 \rightarrow CO_2 + Q. \hspace{3cm} \textbf{13-3}$$

We find that 12 kg of C combine with $2 \times 16 = 32$ kg of O_2 to form 44 kg of CO_2 and that Q is about 4×10^8 J. The 12 kg of C contain the same number of atoms as there are molecules in 32 kg of O_2 and in 44 kg of CO_2; this common number is Avogadro's number $N_0 = 6 \times 10^{26}$. Here N_0 atoms of carbon combine with $2N_0$ atoms of oxygen to form N_0 molecules of CO_2, each molecule of which contains three atoms, (one of C and two of O). Thus the conservation of atoms principle holds for the reaction. Since the mass equivalence of Q is again negligibly small, the conservation of mass principle is also valid.

When iron rusts, oxygen atoms are added to those of the metal, and thus the increase in mass of a rusty metal is explained.

Owing to the small mass equivalence of Q in chemical reactions, the conservation of atoms principle and the conservation of mass principle hold together, as though they were a single law.

13|4 NUCLEAR CHEMISTRY

In 1919 Lord Rutherford performed and explained the first experiment in which *man* changed atoms of one kind of element into those of another. This was the beginning of our current nuclear era. The medieval alchemists had sought to turn base metals into precious ones, and some pretended to have done so, but they never really succeeded. We shall see in Section 13–6 that radioactivity is a *natural* process involving transmutation of one kind of atom, such as radium, into another, such as lead.

Rutherford had observed that radium and associated radioactive materials emitted various radiations among which were high-speed positively charged particles called *alpha particles*. He collected these in an evacuated vessel and proved by spectral analysis of the gas that accumulated that he was collecting *helium*. Thus the alpha particle is a helium nucleus, or a helium atom with its electrons stripped off. These particles served as the bullets in Rutherford's transmutation experiment in 1919. He observed that when alpha particles from a given source passed through air they usually traveled just so far and stopped; that is, they had a definite range in air. Sometimes, however, particles seemed to travel at least four times farther. These long-range particles were identified as protons, or hydrogen nuclei, by (1) the way in which they ionized air and so produced vapor tracks in cloud chambers* (see Fig. 13–2), (2) their long range, and (3) the manner in which they could be deflected by magnetic fields. These protons did not appear when pure oxygen was substituted for air, and they appeared in greater numbers when pure nitrogen was used. In this systematic and scientific manner Rutherford found that when his alpha particles struck nitrogen nuclei some of the latter were transmuted into oxygen nuclei, while protons of high energy and long range were emitted. The reaction may be expressed as follows:

$$_2He^4 + {_7}N^{14} \rightarrow {_8}O^{17} + {_1}H^1 + Q. \qquad \textbf{13–4}$$

This is the standard form for expressing nuclear reactions, so it will be explained in detail.

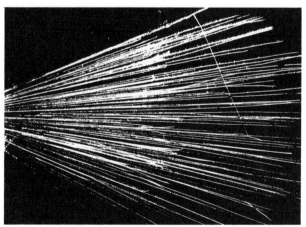

FIG. 13–2
Photograph of α-particle tracks in a cloud chamber. Note the two ranges present here and one example of a nuclear collision and transmutation.

Note that each chemical symbol in Eq. (13–4) has both a subscript and a superscript. The subscript is the *atomic number Z* of the element, or its position number in the periodic table. Helium is the second element, nitrogen the seventh, oxygen the eighth, and the proton is the nucleus of

---

* In cloud chambers we view the tracks of charged particles just as we do the vapor trails of fast airplanes and missiles in the sky. The particles ionize the air and the ions in turn serve as centers of moisture condensation. The water drops formed are large enough to be visible.

TABLE 13–1
Abundance of Stable Isotopes of the Ten Lightest Elements

Element	Isotope	Relative abundance	Element	Isotope	Relative abundance
Hydrogen	H^1	99.985%	Nitrogen	N^{14}	99.6%
	H^2	0.015%		N^{15}	0.4%
Helium	He^3	$10^{-4}\%$	Oxygen	O^{16}	99.76%
	He^4	$\sim$100%		O^{17}	0.04%
Lithium	Li^6	7.5%		O^{18}	0.20%
	Li^7	92.5%	Fluorine	F^{19}	100%
Beryllium	Be^9	100%	Neon	Ne^{20}	90.92%
Boron	B^{10}	18.7%		Ne^{21}	0.26%
	B^{11}	81.3%		Ne^{22}	8.82%
Carbon	C^{12}	98.9%			
	C^{13}	1.1%			

hydrogen, element number one. Rutherford's work on the scattering of alpha particles by nuclei showed that the charge on a nucleus is Ze, where $e = 1.6 \times 10^{-19}$ C, the magnitude of the electronic charge. Therefore, according to the conservation of charge principle, the subscripts should add up to the same value on each side of a nuclear equation. Thus $2 + 7 = 8 + 1$. Since the subscript and the chemical symbol both identify an element, the fact that element number eight must appear on the right-hand side in order to balance subscripts leads us to conclude that the nitrogen nucleus has been changed into one of oxygen.

The superscript is used to represent the *mass number A*, which is defined as the integer nearest the actual mass of the atom when the mass is expressed on the scale C^{12} equals exactly 12 atomic mass units ($C^{12} = 12$ amu).* For most of the chemical elements it has been found that atoms of two or more different mass numbers exist. Thus ordinary oxygen is a mixture of 99.76% O^{16}, 0.04% O^{17}, and 0.20% O^{18}. These three kinds of oxygen atoms are called the stable *isotopes* of oxygen. The stable isotopes of hydrogen are H^1 (99.985%) and H^2 (0.015%); the latter is called "heavy hydrogen" or *deuterium*. The stable isotopes of the ten lightest elements are listed in Table 13–1. The superscripts on each side of a nuclear equation must balance, since otherwise the conservation of mass principle would be grossly violated. The fact that many types of

---

* A slightly different scale that was used until recently took O^{16} to have a mass of exactly 16 units ($O^{16} = 16$ u).

TABLE 13-2
Some Isotopic Masses (C^{12} = 12 amu)

Name	Symbol	Mass, amu*	Name	Symbol	Mass, amu*
Neutron	n^1	1.00867	Boron	B^{11}	11.00930
Hydrogen	H^1	1.00783	Carbon	C^{12}	12.00000
Deuterium	H^2	2.01410	Nitrogen	N^{14}	14.00307
Tritium	H^3	3.01605	Oxygen	O^{16}	15.99491
Helium	He^4	4.00260	Oxygen	O^{17}	16.99907
Lithium	Li^7	7.01600	Uranium	U^{235}	235.0439
Beryllium	Be^9	9.01219			

* On the O^{16} = 16.0000-u scale these values are raised by 0.03%.

nuclear transmutation have been performed since Rutherford's original experiment and that in every case the nuclear equation is one in which subscripts and superscripts balance may be taken as strong proof of the conservation of charge and mass principles.

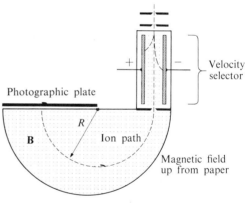

FIG. 13-3
A mass spectrometer. Ions have known charge q. Velocity selector determines their speed v. Where ions strike and activate the photographic plate determines the radius R of their path. The mass m of the ions is then $m = qRv/B$.

Let us now look at the question of mass conservation more closely. Masses of isotopes have been measured very accurately by electromagnetic deflection experiments, as shown in Fig. 13-3. Some typical values are listed in Table 13-2. We see that the masses are not exactly integers. If we add the masses for the isotopes on each side of Eq. (13-4), respectively, we find:

$$He^4 = 4.00260 \qquad O^{17} = 16.99907$$
$$N^{14} = 14.00307 \qquad H^1 = 1.00783$$
$$\overline{Sum = 18.00567} \qquad \overline{Sum = 18.00690}$$

313

Thus the atomic masses on the two sides of the equation fail to balance by

$$18.00690 - 18.00567 = 0.00123 \text{ amu},$$

where amu stands for *atomic mass units*. While this gain in mass is small percentagewise, it is much greater than the possible errors involved in the determination of isotopic masses and so must be accepted as real. Its explanation is based on Einstein's principle of the equivalence of mass and energy, $E = mc^2$, of Section 10–10. In the reaction of Eq. (13–4) the KE (kinetic energy) of the alpha particle (He^4) exceeds that of the O^{17} and H^1 together, so that the reaction is endothermal and Q is negative. We shall see that the mass equivalence of Q turns out experimentally to be equal to minus the gain in mass calculated above. Hence if we count the mass equivalence of the Q term in Eq. (13–4), mass is exactly conserved. The conservation of mass principle is now to be regarded as identical with the conservation of energy principle, or, as we now say, *the conservation of mass-energy principle.*

Example. Find the energy whose mass equivalence is 1 amu so that we may compute Q in Eq. (13–4).

Solution. Since an atomic mass unit and a chemical mass unit are very nearly the same, we may say that 6×10^{26} atoms of C^{12} have a mass of 12 kg. The mass of one atom of C^{12}, which we also call 12 amu, must be $12 \text{ kg}/(6 \times 10^{26})$, so that

$$1 \text{ amu} = \frac{1 \text{ kg}}{6 \times 10^{26}} = 1.66 \times 10^{-27} \text{ kg}.$$

The energy E whose equivalent mass is 1.66×10^{-27} kg is

$$E = mc^2 = (1.66 \times 10^{-27} \text{ kg}) \times (3 \times 10^8 \text{ m/sec})^2 = 1.5 \times 10^{-10} \text{ J}.$$

It is customary to express atomic energies in terms of the small energy unit called the electron-volt, or eV, defined as the energy gained by an electron when accelerated through a potential difference of 1 V. Thus

$$1 \text{ eV} = 1.6 \times 10^{-19} \text{ C} \times 1 \text{ J/C} = 1.6 \times 10^{-19} \text{ J},$$

and the energy equivalent to 1 amu is

$$E = \frac{1.5 \times 10^{-10} \text{ J}}{1.6 \times 10^{-19} \text{ J/eV}} = 930 \times 10^6 \text{ eV} = 930 \text{ MeV},$$

where MeV stands for *million electron-volts*.

We now have for Q in Eq. (13–4)

$$Q = -0.00123 \text{ amu} \times 930 \text{ MeV/amu} = -1.14 \text{ MeV}.$$

When Rutherford used alpha particles with a KE of 7.7 MeV he found that the KE of the H^1 was 6.2 MeV and that of the O^{17} was 0.4 MeV, giving an experimental value for $Q = 6.6 - 7.7 = -1.1$ MeV, which agrees with the value computed according to the conservation of mass-energy principle. Since Q is negative, the initial KE must exceed 1.14 MeV if the reaction is to proceed; actually the alpha particle must have a much greater energy in order to penetrate into the positively charged nitrogen nucleus and trigger the reaction.

13|5 PARTICLE CONSERVATION IN ARTIFICIAL NUCLEAR TRANSMUTATIONS

We have seen that in nuclear reactions the conservation of mass is regarded as part of the conservation of energy principle rather than the conservation of matter principle. As for the latter it is evident that we must again modify our views. The theory of chemical reactions required us to substitute the conservation of atoms principle for the conservation of molecules principle. Now we see that in nuclear reactions not even atoms are conserved! Charge, energy, and *mass number* are conserved; what does the latter represent?

Physicists currently postulate that all nuclei are composed of protons and neutrons. While the proton is positively charged and the neutron is uncharged, these two particles have about the same mass, namely one that is approximately 1840 times that of the electron, or 1 amu. These two particles are thus relatively "heavy" particles; they are also classified together as *nucleons*, or nuclear constituents. Since only the protons contribute to the nuclear charge, each proton having a charge of $+e$, a nucleus of atomic number Z, with a charge of $+Ze$, must contain Z protons. Each proton and neutron adds about 1 amu to the nuclear mass, hence a nucleus whose mass is close to A amu must contain A nucleons in all. Hence,

$Z =$ number of protons in a nucleus,

$A - Z =$ number of neutrons in a nucleus.

We have seen that in a nuclear reaction such as that of Eq. (13–4) the subscripts add up to the same value on each side of the equation and

315

so do the superscripts. In each term the subscript is Z and the super-script is A. Hence we see that the current theory of nuclear structure postulates that in an artificial transmutation

the number of protons remains constant and the number of neutrons remains constant.

This is the third form in which we have stated the conservation of matter principle.

Illustration 1. *Discovery of the neutron*

In 1932 James Chadwick, who had worked with Rutherford, discovered the neutron by bombarding beryllium with alpha particles. This ex-periment had already been tried by M. and Mme. Curie-Joliot in France, who observed that beryllium bombarded with alpha particles yielded a penetrating radiation which, in turn, could knock protons out of matter containing hydrogen. What was the nature of this penetrating radiation? The first guess was that it was similar to the most penetrating x-rays, i.e., electromagnetic radiation of very high frequency. Chadwick pro-ceeded to show experimentally that this radiation did not interact with matter the way such radiation should. However, if he assumed that the new radiation consisted of a stream of neutral particles ("neutrons") with a mass near that of the proton, then all of his results could be explained. For example, a head-on collision of a neutron with a proton of the same mass should result, according to Newton's laws, in the complete transfer of KE from neutron to proton, just as in the head-on collision of two billiard balls of equal mass. The impact of a neutron with a nitrogen nucleus of fourteen times its mass should result in little transfer of KE, just as when a pea strikes a tennis ball. As predicted by the neutron hypothesis, Chadwick found that the penetrating radiation transferred much more energy to protons than to nitrogen nuclei; he computed the neutron's mass to be about 1 amu.

The alpha particle–beryllium reaction thus is the following:

$$_2\text{He}^4 + {_4}\text{Be}^9 \rightarrow {_6}\text{C}^{12} + {_0}\text{n}^1 + Q, \qquad \textbf{13-5}$$

where n is the symbol for the neutron. Here we see that the alpha particle adds 2 protons and 2 neutrons to a Be nucleus that is composed of 4 protons and 5 neutrons (the only stable isotope of Be). The resulting compound nucleus of 6 protons and 7 neutrons is unstable and a neutron is ejected, leaving a C^{12} nucleus. Figure 13–4 shows the reaction pictorially.

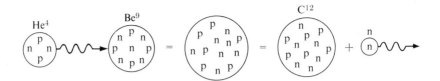

FIG. 13–4. Conservation of neutrons and protons in a nuclear transmutation.

Illustration 2. *First transmutation with accelerated particles*

In the reaction

$$_1H^1 + {}_3Li^7 \rightarrow {}_2He^4 + {}_2He^4 + Q, \qquad \qquad \textbf{13-6}$$

the product nuclei are both alpha particles. Since $Q = +17.2$ MeV, Sir John Cockcroft of the Cavendish Laboratory in Cambridge, England, was able to obtain this reaction by accelerating protons to rather modest energies (about 0.3 MeV) and firing them at a lithium target; he and Walton, his colleague, showed that pairs of oppositely directed, high-energy alpha particles were produced, as shown in Fig. 13–5. In this reaction the constant number of protons is 4 and the constant number of neutrons is also 4. The combination of 2 protons plus 2 neutrons, i.e., the alpha particle, is evidently very stable.

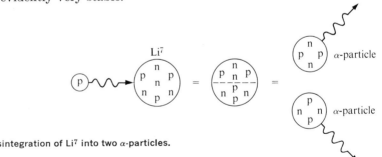

FIG. 13–5
Cockcroft and Walton's disintegration of Li^7 into two α-particles.

13|6 PARTICLE CONSERVATION IN RADIOACTIVE TRANSFORMATIONS

Radioactivity was first discovered in 1896 by Henri Becquerel. Here again is an example of how an experiment undertaken to explore one phenomenon may lead an alert and trained observer to the discovery of something quite new and different.

Becquerel was initially interested in the radiation emitted by certain substances after they have been exposed to light. We say that such substances are *phosphorescent;* after exposure to light they will glow in the dark. Becquerel wondered whether there was any connection between phosphorescent radiations and the x-rays that the German physicist Roentgen had discovered only a few weeks before. By chance Becquerel left a phosphorescent uranium compound that he had *not yet* exposed to light next to a photographic plate that was wrapped in heavy black paper, and by chance he later developed the plate and found it blackened as though it had been greatly exposed to radiation of some kind. Becquerel was quick to trace the radiation to its source, namely the uranium compound.

The most active ingredient in uranium ore was painstakingly tracked down by Mme. Curie, who in 1898 isolated the element radium and showed that it was a much more intense source of radiation than uranium.

We now know that nuclei of uranium, radium, and other heavy elements may undergo spontaneous transmutation with the emission of three kinds of rays, α, β, and γ (alpha, beta, and gamma). We have seen that the α-particle is a helium nucleus, the β-particle is an electron, and γ-rays ("gamma rays") are very penetrating electromagnetic radiation. A γ-ray photon may possess an energy of several MeV; this makes it destructive to living tissue and capable of inducing gene mutations in cell chromosomes.*

Since 1934 it has been found possible to produce artificially radioactive isotopes of the common lighter elements. This may be accomplished through various nuclear transmutations, but the most effective methods are (1) to bombard an element with neutrons or (2) to induce nuclei of a heavy element to *fission*, i.e., to split in two. Typical reactions leading to a radioactive product are the following:

$$_{13}\mathrm{Al}^{27} + {_0}\mathrm{n}^1 \rightarrow {_{13}}\mathrm{Al}^{28} + Q, \tag{13–7}$$

$$_{7}\mathrm{N}^{14} + {_0}\mathrm{n}^1 \rightarrow {_6}\mathrm{C}^{14} + {_1}\mathrm{H}^1 + Q, \tag{13–8}$$

$$_{92}\mathrm{U}^{235} + {_0}\mathrm{n}^1 \rightarrow {_{38}}\mathrm{Sr}^{90} + {_{54}}\mathrm{Xe}^{136} + 10\,{_0}\mathrm{n}^1 + Q. \tag{13–9}$$

---

* While there is no threshold of intensity below which γ-rays are completely harmless, the risk of danger from weak doses is small. Although a dose of 400 roentgens (about 4 J of radiant energy absorbed per kilogram of body) in a short period of time is regarded as lethal, the normal background radiation is only 0.1 or 0.2 roentgen per year.

In (13–7), Al^{27}, the only stable isotope of aluminum, is changed into Al^{28}, which is radioactive. In (13–8) radioactive C^{14} is formed. The stable isotopes of carbon are C^{12} and C^{13}; C^{10}, C^{11}, and C^{14} are radioactive (semistable) and all other possible carbon nuclei are completely unstable. Equation (13–9) represents the fission process in which radioactive strontium and xenon are formed. There are other possible ways in which U^{235} may fission (see Section 14–2), but (13–9) is a fairly likely one. Since "A-bombs" explode through the fission process, and since "H-bombs" employ A-bombs to build up the high temperature required to explode an H-bomb, both types of bombs release radioactive "fallout" containing such products as Sr^{90} and Xe^{136}.

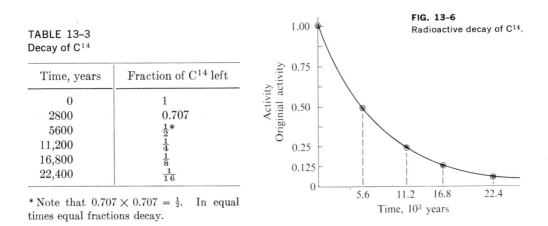

FIG. 13–6
Radioactive decay of C^{14}.

TABLE 13–3
Decay of C^{14}

Time, years	Fraction of C^{14} left
0	1
2800	0.707
5600	$\frac{1}{2}$*
11,200	$\frac{1}{4}$
16,800	$\frac{1}{8}$
22,400	$\frac{1}{16}$

* Note that $0.707 \times 0.707 = \frac{1}{2}$. In equal times equal fractions decay.

An important property of radioactivity is that the rate of decay of a given isotope is characteristic of that isotope and is unaffected by the physical environment (temperature, pressure, state, etc.) or by chemical combination. The decay of a pure sample of some radioactive isotope is exponential, which means that if half decays in 5600 years, as happens with C^{14}, then half of what is left after 5600 years will decay in another 5600 years, and so on. We say that C^{14} has a *half-life* of 5600 years. Its activity as a function of time is shown in Fig. 13–6 and the fraction remaining at various times is indicated in Table 13–3. The fact that the ratio of C^{14} to stable carbon is fairly constant in living matter, but in dead matter decays with the half-life of 5600 years, has made C^{14} a useful tool in dating archeological remains of matter that was once living.

319

Let us now examine the process, known as *beta decay*, by which nuclei such as C^{14} and C^{11} change into stable nuclei. For isotopes such as C^{14}, in which the number of neutrons is apparently too great for stability (remember that C^{12} and C^{13} are stable), decay occurs by a process in which a neutron near the edge of the nucleus is, in effect, changed into a proton plus an electron plus a neutrino; the proton remains in the nucleus and the electron is emitted as a β-ray, accompanied by the neutrino. The latter is a particle with no charge and no appreciable rest mass, but it carries away energy and momentum. This explanation was proposed by Enrico Fermi (1901–1954), the great Italian-American physicist who first made a nuclear chain reaction work. Recall that in Chapter 5 we saw that new forms of energy have been postulated whenever new discoveries seem at first to suggest a violation of the conservation of energy principle. Thus in β-decay one finds an apparent loss of energy in the usual forms, so we say that the neutrino carries away the unaccounted-for energy. The neutrino and its energy can rarely be recaptured, since the neutrino is chargeless (not subject to electromagnetic forces), massless (not subject to gravitational or inertial reactions), and not a photon of electromagnetic radiation. Fermi worked out a self-consistent theory which has "grown gracefully"; the neutrino's existence is now firmly supported by direct as well as by indirect evidence.

Letting $_{-1}\beta^0$ represent the electron and $_0\bar{\nu}^0$ the neutrino, we have for the decay of C^{14}

$$_6C^{14} \rightarrow {_7N^{14}} + {_{-1}\beta^0} + {_0\bar{\nu}^0} + Q, \qquad \textbf{13-10}$$

where the product nucleus, N^{14}, is the common stable isotope of nitrogen. Note that the mass number of the electron is zero, since its actual mass is only 0.00055 amu, for which the nearest integer is zero. The conservation of subscripts and of superscripts holds in Eq. (13–10), but the pictorial representation of the transmutation shown in Fig. 13–7 makes it obvious that neither the number of protons nor the number of neutrons remains constant. There are 6 protons and 8 neutrons at the start, and there are 7 protons and 7 neutrons at the end. Hence we must once again modify our conservation of matter principle and state that

in β-decay the number of nucleons, or protons plus neutrons, remains constant.

The C^{11} nucleus is unstable, presumably because it contains too *few* neutrons. It undergoes a decay process in which a proton is changed into a neutron, a positron, and a neutrino. The *positron* is the positive counterpart of the ordinary negative electron. Both particles have the same mass. The positron was first discovered in the cosmic radiation in 1932. It is a rare particle, whereas the electron is a constituent of all our matter. We may represent the positron by the symbol $_{+1}\beta^0$. The decay of C^{11} is then given by the equation

$$_6C^{11} \rightarrow {}_5B^{11} + {}_{+1}\beta^0 + {}_0\nu^0 + Q, \qquad \text{13-11}$$

where $_0\nu^0$ is the neutrino. (The difference between $\bar{\nu}$ and ν will be explainly shortly.) The boron isotope B^{11} is stable.

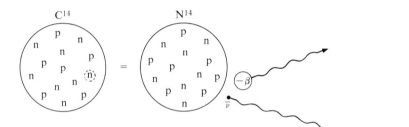

FIG. 13-7
In β-decay of a heavy isotope a neutron is changed into a proton plus an electron plus a (anti) neutrino.

In β-decay charge is conserved and (with the postulate of the neutrino) mass-energy is conserved, but the conservation of matter seems to apply only with respect to the number of heavy particles, or nucleons. Electrons, positrons, and neutrinos appear to be created. We investigate this further in the next section.

13|7 PAIR ANNIHILATION AND PAIR PRODUCTION

In the example at the end of Section 10–10, reference was made to the fact that when a positron meets an electron the two particles annihilate one another. Their charges of course neutralize each other ($+e - e = 0$) and so in a sense does their matter. The combined mass of the two particles is 18.2×10^{-31} kg, or 0.0011 amu, and the equivalent energy is

0.0011 amu $\times$ 930 MeV/amu $= 1.02$ MeV.

It has been observed that when positrons from a source such as C^{11} are absorbed by matter, pairs of γ-ray photons, each bearing the energy 0.51 MeV, are emitted in opposite directions. This is strong evidence for postulating that the reaction is that shown in Fig. 13–8, namely

$$_{-1}\beta^0 + {_{+1}}\beta^0 \rightarrow 2\ \gamma\text{-ray photons,} \qquad \textbf{13-12}$$

and that in this process the conservation of mass-energy as well as of charge hold true. However, electrons, positrons, and photons are not individually conserved.

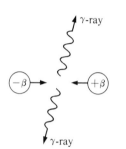

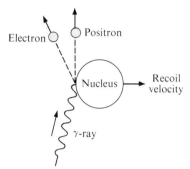

FIG. 13–8. Annihilation of electron and positron.

FIG. 13–9. Pair production occurs only in the presence of matter.

The reverse process, pair production, may occur when a γ-ray photon bearing an energy in excess of 1.02 MeV encounters matter. The mass associated with the photon materializes as that of the created electron and positron. The conservation of momentum principle (a corollary of Newton's laws) requires that pair production occur in the presence of some particle of matter that may absorb momentum (see Fig. 13–9).

It was predicted that just as the positron is the positively charged counterpart of the electron, there should be the negative counterpart of the positive proton. This predicted particle, called the *antiproton*, should be capable of annihilating a proton and it should make its appearance, along with a proton, through a process of pair production. Since the process

$$Q\ (\text{energy}) \rightarrow \text{proton} + \text{antiproton} \qquad \textbf{13-13}$$

involves the materialization of about 2 amu (3.3×10^{-27} kg) of mass, Q must exceed

$$1860\ \text{MeV} = 1.86 \times 10^9\ \text{eV.}$$

This high energy of around 2 BeV (billion electron-volts) has been given to particles in the largest atomic accelerators, and antiprotons have been created by such machines since 1955.

Pair production of heavy particles results in altering the total number of protons plus neutrons in our world. Once again we must modify our conservation of matter principle! This time we shall see that it may be "saved" by introducing the new concept of *antimatter*. The "anti-**P** particle" is defined as that particle which (1) can annihilate a "**P**-particle" and (2) is created in pair production along with a "**P**-particle." Thus the positron is the electron's antiparticle, the antiproton is the proton's, and so on. The $\bar{\nu}$ of Eq. (13–10) is the antiparticle of the ν in Eq. (13–11); $\bar{\nu}$ is often called the *antineutrino*, but ν and $\bar{\nu}$ are alike in charge (they are both uncharged). The neutrino that appears in the β-decay of a heavy isotope like C^{14} is always the antineutrino, whereas the neutrino that appears in the β-decay of a light isotope such as C^{11} is the neutrino.

Let us start to count the particles in a closed system and as we do so let us count "plus one" for every proton, neutron, electron, and neutrino and "minus one" for every antiproton, antineutron, positron, and antineutrino. This is the sort of thing we would do when counting up positive and negative charges to find the total net charge in a region. When we count an antiparticle as minus a particle we do not mean that it is negative with respect to charge or mass; we only imply that it is negative in our counting system. If we apply this system of counting to pair production and annihilation, then we may state that in pair annihilation and pair production, as well as in the radioactivity and nuclear transmutation processes, *the net number of particles in our world remains constant*, provided that photons are not counted as particles. This is still another modification of the conservation of matter principle.

Example 1. Consider the process of pair production, as given by Eq. (13–13). The system starts with a particle count of zero and ends with a count of $1 - 1 = 0$.

Example 2. In pair annihilation, Eq. (13–12), the system starts with a particle count of $1 - 1 = 0$ and, since we do not count γ-ray photons, the count at the end is zero.

Example 3. In the β-decay of C^{11}, Eq. (13–11), the system starts with a count of 11 and ends with one of $11 - 1 + 1 = 11$. Here the positron counts -1 and neutrino $+1$.

Example 4. In the β-decay of C^{14}, Eq. (13–10), the count remains at 14 because the electron counts $+1$ and the antineutrino -1.

Example 5. In artificial transmutation processes antiparticles are not involved; hence the number of ordinary particles is conserved.

13|8 CONSERVATION OF ELEMENTARY PARTICLES. FUNDAMENTAL LAW XV

The reader may have read from time to time about the various new particles of matter that physicists have been discovering. Mention has been made in this chapter of the discovery of the neutron, the positron, the antiproton, the neutrino, and the antineutrino, while earlier in the book the electron and proton were introduced. These are all called *elementary particles* to indicate that they are not, as far as we know, compounded out of more fundamental particles. This may eventually prove to be a poor assumption; remember that the atom was once regarded as indivisible and now we postulate a very complex structure for an atom of a heavy element! Be that as it may, we find that at present physicists have classified the various elementary particles as follows.

a | Photons

Photons, or light quanta, are the bundles of energy in which electromagnetic waves are emitted and absorbed. A photon travels with the speed of light and it ceases to exist when stopped. It has mass associated with its energy, but no rest mass. Its energy is related to the frequency f of the waves through the postulate of Planck,

$$E = hf.$$

We have seen that the photon is not counted as a particle of matter.

b | Leptons

These light-weight particles include the following:

$$\left. \begin{array}{l} \nu_e, \nu_\mu \\ \bar{\nu}_e, \bar{\nu}_\mu \end{array} \right\} \text{ neutrinos*} \qquad\qquad \text{mass} \sim 0,$$

$$\left. \begin{array}{l} e^- = -\beta = \text{electron} \\ e^+ = +\beta = \text{positron} \end{array} \right\} \quad \text{mass} = m_0 = 9.1 \times 10^{-31} \text{ kg},$$

$$\left. \begin{array}{l} \mu^- \\ \mu^+ \end{array} \right\} \text{ muons} \qquad\qquad \text{mass} = 207\, m_0.$$

The muons form a particle-antiparticle pair.

---

* Recent evidence indicates that the neutrinos ν_e and $\bar{\nu}_e$ associated with electrons and positrons react differently with nucleons than do the neutrinos ν_μ and $\bar{\nu}_\mu$ associated with muons.

c | Mesons

These intermediate-weight particles include

$$\left.\begin{array}{l}\pi^- \\ \pi^+ \\ \pi^0\end{array}\right\} \text{pions} \qquad\qquad \text{mass} = 270\ m_0,$$

K-particles, or heavy mesons mass $\sim 1000\ m_0$.

The name *meson* refers to the fact that a meson particle has a mass whose value lies *between* that of a lepton and that of the proton or neutron.

d | Baryons

These heavy particles include

$$\left.\begin{array}{ll}\text{p}^+ = & \text{proton} \\ \text{p}^- = & \text{antiproton} \\ \text{n} = & \text{neutron} \\ \bar{\text{n}} = & \text{antineutron}\end{array}\right\} \text{mass} \sim 1840\ m_0,$$

hyperons $\qquad\qquad$ mass $\sim 2200\ m_0$ or greater.

An important characteristic of all the above particles, except the photon, the neutrinos, the electron and positron, and the proton and antiproton, is their *instability*. Mesons and hyperons decay with very short half-lives into lighter particles. Even the neutron decays with a half-life of 12.5 minutes when it is free, although within a nucleus it appears to be stable. The various modes of decay are most complex and not yet completely understood, but we shall study a few in order to see in what manner the conservation of matter principle must be stated for elementary particle transformations.

The muons decay as follows:

$$\mu^- \rightarrow e^- + \bar{\nu}_e + \nu_\mu + Q, \qquad \mu^+ \rightarrow e^+ + \nu_e + \bar{\nu}_\mu + Q. \qquad \textbf{13-14}$$

To ensure particle conservation in these two processes we count "minus one" for the μ^+ as well as for the e^+, and "plus one" for the μ^- and the e^-. In other words, the negative muon, the electron, and the proton are all particles of the same kind of matter, the kind that predominates in our own world. The μ^+ and e^+ are both particles of antimatter.

The pions decay as follows:

$$\pi^- \rightarrow \mu^- + \bar{\nu}_\mu + Q, \qquad \pi^+ \rightarrow \mu^+ + \nu_\mu + Q, \qquad \pi^0 \rightarrow \gamma + \gamma + Q.$$

$$\textbf{13-15}$$

[Here the ν_μ and $\bar{\nu}_\mu$ react differently with matter than do the ν_e and $\bar{\nu}_e$ in Eqs. (13–14).] We have ruled that the μ^+ and $\bar{\nu}$ are particles of antimatter and the μ^- and ν are particles of matter. The γ-ray photon is not counted. Then the count for the right-hand side of each of these equations is 0. In order to maintain the particle count constant we must say that *we will not count the pion!* The decay of K-mesons indicates that they must not be counted either. Hence we see that the conservation of particles does not apply to mesons or photons.

The decay of the neutron is given by

$$_0n^1 \rightarrow {}_1p^1 + {}_{-1}e^0 + \bar{\nu}_e + Q, \qquad\qquad \text{13-16}$$

where $_1p^1$ stands for the proton and $_{-1}e^0$ for the electron. Here the particle count remains constant and equal to 1.

The subject of hyperon transformations is a most complex one. The following is a typical decay process:

$$\text{hyperon} \rightarrow \text{proton} + \text{pion}. \qquad\qquad \text{13-17}$$

Here again we must apply the rule above, namely not to count pions.

The general particle conservation law, which stated that *the number of leptons plus baryons of our matter minus the number of leptons and baryons of antimatter appears to remain constant in any transformation*, permits processes that so far have not been observed. Consider, for example, the process $\mu^- \rightarrow e^- + \gamma$. Since the γ-ray photon is not counted, this is an allowed process according to the above postulate, yet it has not been discovered, as it easily would be if it occurred. Current feeling is that any process allowed by conservation laws *should* occur. In other words, conservation laws are regarded as both *restrictive* and *permissive*.

The present status of the conservation of the number of particles in our world is as follows.

The general conservation law stated above has been replaced by a more restrictive trio of particle conservation laws. Consider the e^-, e^+, ν_e, and $\bar{\nu}_e$ as one "family," the μ^-, μ^+, ν_μ, and $\bar{\nu}_\mu$ as a second family, and the baryons as a third. In each family count particles of matter as positive and those of antimatter as negative to determine the net number. Then it appears at present that in any transformation

1. *The electron-family number remains constant.*

2. *The muon-family number remains constant.*

3. *The baryon-family number remains constant.*

Here we have the latest version of the *conservation of matter principle,* our *Fundamental Law XV.*

The evolution of this principle is a good illustration of the way in which physicists modify their views, introduce new concepts such as that of antimatter and the neutrino, and yet hold firm to a basic postulate (conservation of matter in some form) that has stood them in good stead through the years. One also tends to acquire an intuitive belief in what seems to be a general rule; why abandon this belief because of an apparent exception in some esoteric field, especially when modification of the principle is the most fruitful procedure?

PROBLEMS

1. Ten grams of salt are dissolved in 100 gm of water in an open bowl. The next day the mass of the solution is found to be 95 gm. Explain, stating any postulate assumed.

2. How many kilograms of oxygen are chemically bound up in 10 kg of quartz sand (SiO_2)? (The atomic weight of silicon is 28.)

3. If 100 gm of Fe (iron, atomic weight = 55.85) are oxidized into Fe_2O_3, what will the change in mass be?

4. Compute the energy in electron-volts released when one carbon atom is oxidized as in Eq. (13–3). How much is this energy per nucleon of carbon? Per nucleon of CO_2?

5. Complete the following equations and identify the missing term by name.

    a) $Li^6 + n^1 \rightarrow H^3 +$?   b) $H^2 \rightarrow H^1 +$?

    c) $Li^6 + H^2 \rightarrow Li^7 +$?   d) $H^2 + H^2 \rightarrow H^3 +$?

    e) $H^2 + H^2 \rightarrow He^3 +$?   f) $H^2 + H^3 \rightarrow He^4 +$?

6. Compute Q for the process $He^4 \rightarrow 2H^1 + 2n^1 + Q$, using Table 13–2. Why is $-Q$ called the "binding energy"? What is this binding energy per nucleon of He^4?

7. Compute the binding energy per nucleon for (a) C^{12}, (b) O^{16}.

8. Compute Q for process (f) of Problem 5.

9. Compute the energy released when a free neutron decays.

10. The stable isotopes of nitrogen are N^{14} and N^{15}. Write the equations for the decay of N^{13}, N^{16}, and N^{17}.

11.   The half-life of I^{131} is 8 days. In what time will the activity decrease to (a) 25%, (b) about 1%, (c) about 0.1% of its initial value?

12.   If a radioactive sample loses 20% of its activity in 10 hours, what percent will it lose in (a) 20 hours, (b) 40 hours?

13.   a)   If a neutral hyperon decays into a proton and a pion, should the neutral hyperon be regarded as a particle of matter, or as a particle of antimatter?

b)   If the pion decays into a muon and the muon into an electron, list all of the final particles obtained from the decay of the neutral hyperon.

c)   If these final particles are counted according to the rules given, what is the net particle count?

Experiment 20   EXPONENTIAL DECAY
Relaxation Phenomena

Object: To study the physical conditions leading to exponential decay and the mathematical properties of a decay curve.

Theory: As physical systems approach a final state of equilibrium they "relax," as it were.   Let some parameter (variable) y represent the departure of a system from equilibrium at the time t; the rate at which y is decreasing at that instant is $-\Delta y/\Delta t$, where Δy represents the *increase* in y during the change Δt in t.   Now it is a matter of experimental observation that in many relaxation phenomena the final state is approached more and more slowly the nearer the system gets to this state and that the quantitative relationship is

$$-\Delta y/\Delta t = ky, \tag{1}$$

where k is independent of t and y and is called the *decay constant*.   Under these circumstances the decay is always exponential, that is, it is given by the relationship

$$y = y_0 e^{-kt}, \tag{2}$$

in which y_0 is the original value of y (for which $t = 0$). If y is plotted against t, the curve will resemble that in Fig. 13–6.   The greater k, the steeper the curve and the faster the decay.

What distinguishes an exponential curve from other down-sloping curves is that the variable plotted vertically (y) decreases by the same frac-

328

tion or percent every time the variable plotted horizontally (here t) increases by the same numerical amount. Thus if y decreases to half its value in the time $t = T$, it will decrease by half again in another interval T, or in the time $t = 2T$ it will decrease to one quarter of its initial value. T is then called the *half-life*. But one can also say that if y decreases from y_0 to $0.8\,y_0$ in a time T^1, it will decrease from y_0 to $0.8 \times 0.8\,y_0 = 0.64\,y_0$ in a time $2T^1$, etc.

If one plots the log of y against t one should get a straight line if Eq. (2) holds. This is another sensitive test of an exponential relationship.

Procedure: There are many experimental examples of exponential decay; the choice must depend on the equipment available. It is recommended that you choose one of the following experiments and for it plot the decay curve, find the half-life, and apply the exponential tests.

Experiment 1. Measure with a Geiger counter the initial activity y_0 of a sample of radioactive iodine-131. Remeasure the activity at 1-day intervals for 2 or 3 weeks.

Experiment 2. Heat a thimble of water to about 50°C and place it in a large bath of water at room temperature. Here y is the temperature difference between the warm water and the bath. Measure y at 1-minute intervals, starting with $y_0 = 20$°C and continuing until $y = 5$°C or less. Exponential decay would mean that the rate of cooling of a warm body is proportional to the difference in temperature between the body and its surroundings. Newton first propounded this as an empirical law; see how closely you agree with Newton.

Experiment 3. Observe the absorption of beta rays by successive layers of cardboard. Here the thickness of absorbing material plays the role of t.

Experiment 4. Observe the decreasing amplitude of an underdamped oscillating system, say a galvanometer coil.

References

FORD, K. W., *The World of Elementary Particles*, Blaisdell, 1963.

FRISCH, D. H., AND
THORNDIKE, A. M., *Elementary Particles*, Van Nostrand, 1964.

GAMOW, GEORGE, *Mr. Tompkins Explores the Atom*, Cambridge University Press, 1958.

GAMOW, GEORGE, *The Atom and Its Nucleus*, Prentice-Hall, 1961.

PARK, D., *Contemporary Physics*, Harcourt, Brace and World, 1964.

14

14|1 INTRODUCTION Near the close of the 19th century it was felt by a number of physicists that all of the fundamental laws of nature had been discovered and that further work would consist mainly of measuring some quantity to another decimal place. The course of events turned out to be quite different from this prediction. The decade 1895–1905 saw the discovery of x-rays, radioactivity, the electron, the quantum principle, and the relativity principle. This was indeed a fruitful period and the explanation is that in this decade physicists first began to explore a new realm of the physical world, namely inside the atom. When research is carried on in completely new areas the chances of discovering new fundamental principles are greatly improved.

Nuclear Forces and the Search for Additional Principles

At the present time the frontiers of investigation in physics have been pushed far out beyond the realm of our everyday lives. Current fundamental research generally requires increasingly advanced training on the part of the physicist and more sophisticated and expensive equipment with which to work. The trend is toward extremes, such as investigation at very low or very high temperatures, in the very small realm of the atomic nucleus or the huge one we call the universe.

There is a feeling (which could be wrong) that experiments at very low temperatures will probably not reveal any new fundamental principles. Discoveries in solid state research also seem likely to be mostly of technological value. High temperatures, such as ten million degrees or more, involve high energies, and with particles of high energy one may explore the atomic nucleus. Nuclear physics in turn is closely related to the elementary particles and the interactions between these particles. Here in nuclear-particle physics we feel sure that principles are involved that are not yet completely understood, such as (1) the law(s) of nuclear forces and (2) the principle which explains why elementary particles of certain masses, charges, etc., and not other masses, charges, etc., have been discovered in nature.

Considerable thought of a philosophical nature is also being given to the symmetry properties of our physical world and their relation to the various conservation principles. We shall touch upon these after first reviewing our current knowledge of nuclear forces.

14|2 NUCLEAR ENERGY

We saw in the last chapter that when nuclear particles are rearranged the energy change is usually great enough to involve a measurable change in mass. This amounts to saying that nuclear energy changes are much greater than those resulting from the rearrangement of atoms and their orbital electrons in chemical transformations. Let us compare the energy released per kilogram and per nucleon in a typical chemical combustion process with that emitted per kilogram and per nucleon in the nuclear processes of fission and fusion.

a | Combustion

The burning of wood and coal chiefly involves the process

$$C + O_2 \rightarrow CO_2 + Q, \qquad\qquad \textbf{14-1}$$

where $Q = 4 \times 10^8$ J/kg-mole of C, O_2, and CO_2. The atomic weight of C is 12 and that of O is 16. Thus from 12 kg of C and 32 kg of O_2 we get 4×10^8 J of energy. This amounts to 3.3×10^7 J/kg of C. We may regard the carbon as the fuel which must be purchased; the oxygen is ordinarily freely available in our atmosphere.

While it is more convenient in atomic physics to use the electron-volt as the unit of energy rather than the much larger joule, we find that in engineering a still bigger unit called the *kilowatt-hour* is preferred. By definition

$$1 \text{ kilowatt-hour} = 1 \text{ kWh} = 1000 \text{ watt-hr} = 1000 \times 3600 \text{ W-sec}$$
$$= 3.6 \times 10^6 \text{ J.}$$

Thus the energy released in the burning of 1 kg of carbon amounts to

$$\frac{Q}{\text{mass of C}} = \frac{4 \times 10^8 \text{ J}}{12 \text{ kg}} = \frac{10^8 \text{ J}}{3 \text{ kg} \times 3.6 \times 10^6 \text{ J/kWh}}$$
$$\doteqdot 10 \text{ kWh/kg of C.}$$

There are 6×10^{26} atoms in 12 kg of C, and one atom of C contains 12 nucleons. Thus the energy Q is released by an amount of matter that contains 72×10^{26} nucleons. Hence

$$\frac{Q}{\text{No. of nucleons}} = \frac{4 \times 10^8 \text{ J}}{72 \times 10^{26} \text{ nucleons} \times 1.6 \times 10^{-19} \text{ J/eV}}$$

$$= 0.35 \text{ eV/nucleon}.$$

b | Fission

The fission process was mentioned briefly in Section 13–6. It derives its name from the fact that it involves the *splitting in two* of a heavy nucleus (see Fig. 14–1) rather than the transformation of a nucleus into another one of about the same mass.

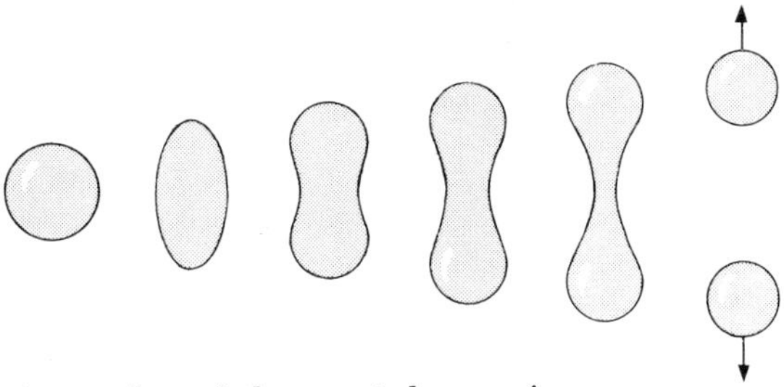

FIG. 14–1
Schematic representation of the fission of a heavy nucleus.

The story of the discovery of fission is another of the great dramas in the history of physics. As early as 1934 Enrico Fermi had conducted experiments in which various nuclei were bombarded with neutrons and radioactive products resulted. Fermi thought, in the case of the bombardment of uranium, that he had produced transuranic (beyond uranium) elements. We now know that this was partly true, but what Fermi did not realize was that he had also caused the fission of some uranium nuclei.

Two German physical chemists, O. Hahn and F. Strassmann, published on January 6, 1939, a paper in which they described the puzzling results of their work on the chemical identification of the products formed when uranium was bombarded with neutrons. They wrote that the evidence pointed to the formation of nuclei much lighter than uranium, but this seemed incredible to them in view of the fact that all previous experiments in nuclear transmutation had involved only a change of 1 or 2 in the atomic number of the affected nucleus. However, Lise Meitner, a former colleague of Hahn, and Otto Frisch did believe in the fission interpretation of Hahn's and Strassmann's results and they so informed Neils Bohr, who, in turn, announced their conclusions at a meeting of the American Physical Society in Washington, D.C., on January 26, 1939. Almost overnight American physicists undertook, in different laboratories, confirmatory experiments.

The importance of the discovery lies in the fact that the fission of a uranium nucleus requires an impinging neutron, while as a result of the fission one obtains not only the two product nuclei plus a relatively large release of energy, but *several neutrons* as well. It was easy to see that if one neutron could produce two, two might produce four, four produce eight, and so on (Fig. 14–2). This would be a *chain reaction*, making possible the release of nuclear energy on a large scale. The atomic bomb and the many nuclear power plants on land and in submarines testify to man's success in overcoming the various technical problems that had to be surmounted. Thus January 1939 marks the beginning of the age of nuclear energy on our earth.

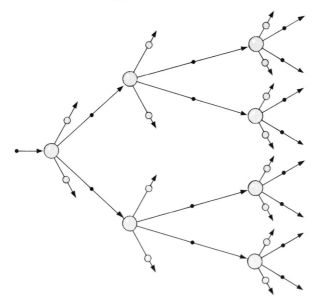

FIG. 14–2
Simplified sketch of a chain reaction in U²³⁵.

In the typical fission reaction

$$U^{235} + n^1 \rightarrow Ba^{141} + Kr^{92} + 3n^1 + Q, \hspace{2cm} \textbf{14–2}$$

Q may be calculated from our knowledge of nuclear masses and the principle of the conservation of mass-energy. It is found that Q is about 210 MeV/U-nucleus. Since 930 MeV are equivalent to 1 amu (see Example in Section 13–4), we see that here Q is equivalent to $\frac{210}{930}$ amu $=$ 0.225 amu, which is about 0.1% of the mass (235 amu) of the U nucleus. Hence the fission of 1 kg of U is accompanied by the release of energy equivalent to, or associated with, $0.1\% \times 1$ kg $= 10^{-3}$ kg of mass.

From Einstein's relation $E = mc^2$, we have

$$\frac{Q}{\text{mass of U}} = (10^{-3} \text{ kg/kg}) \times (3 \times 10^8 \text{ m/sec})^2 = 9 \times 10^{13} \text{ J/kg}$$

$$= \frac{9 \times 10^{13} \text{ J}}{1 \text{ kg} \times 3.6 \times 10^6 \text{ J/kWh}} = 25 \times 10^6 \text{ kWh/kg}.$$

This number is two and one-half million times greater than in the case of the combustion of carbon!

A U^{235} nucleus contains 235 nucleons, so that

$$\frac{Q}{\text{No. of nucleons}} = \frac{210 \text{ MeV}}{235 \text{ nucleons}} = 0.9 \text{ MeV/nucleon}.$$

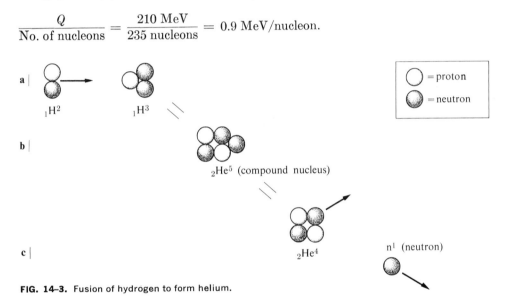

FIG. 14–3. Fusion of hydrogen to form helium.

It is this great concentration of energy in fission fuel that gives such a fuel its great importance. Nuclear reactors use the fission process in a controlled manner and A-bombs, in an explosive, uncontrolled reaction.

c | Fusion

The direct combination of two hydrogen atoms and two neutrons should give a helium atom and, according to the measured masses of H^1, n^1, and He^4, a Q of 28 MeV. This process of *combining* lighter particles to form a heavier nucleus is called *fusion*. Actually it is not practical to try to get four particles to collide simultaneously, but in the following fusion process (Fig. 14–3) only two particles need collide:

$$H^2 + H^3 \rightarrow He^4 + n^1 + Q,$$

14-3   **335**

where Q is 17.6 MeV/He-nucleus. Here the H^2 is the nucleus of the heavy hydrogen atom found in a concentration of 0.015% in all of our water. The H^3 may be obtained from the process represented by

$$H^2 + H^2 \rightarrow H^3 + H^1. \qquad\qquad \textbf{14–4}$$

The H^3 is radioactive, but as its half-life is 12 years it is feasible to stock-pile it. Since 17.6 MeV are equivalent to 17.6/930 amu = 0.019 amu, which is 0.38% of the mass (5 amu) of the combining nuclei, we may say that here the energy released per kg is that equivalent to, or associated with, 0.0038 kg. Thus from $E = mc^2$ we have

$$\frac{Q}{\text{mass of H}} = (0.0038 \text{ kg/kg}) \times (3 \times 10^8 \text{ m/sec})^2$$

$$= \frac{3.42 \times 10^{14} \text{ J}}{1 \text{ kg} \times 3.6 \times 10^6 \text{ J/kWh}}$$

$$= 95 \times 10^6 \text{ kWh/kg}.$$

This represents nearly four times the concentration of energy in fission fuel. Since H^2 and H^3 together contain 5 nucleons, we have

$$\frac{Q}{\text{No. of nucleons}} = \frac{17.6 \text{ MeV}}{5 \text{ nucleons}}$$

$$= 3.5 \text{ MeV/nucleon}.$$

The reaction in Eq. (14–3) will take place only if the colliding nuclei have high initial speeds; otherwise they will not, as it were, stick together. As two nuclei approach each other the Coulomb force of repulsion between their like (positive) charges predominates until they are sufficiently close for nuclear attractive forces (see the following section) to take control. To ensure collisions at high speeds one must heat the hydrogen sufficiently so that a significant fraction of the particles have, according to the kinetic theory, the kinetic energies associated with the required speeds. Since it turns out that the fusion reaction will only take place at temperatures of many millions of degrees, it is termed a *thermonuclear reaction*. In the H-bomb the required temperature is acquired through the explosion of an A-bomb and then the fusion process is utilized in an uncontrolled manner. The controlled use of the fusion process would make it possible to utilize sea water as a source of energy, a source available in abundance to every nation. The great experimental problem of today is to discover how to build a practical device in which the fusion process is made to proceed in a controlled manner. One attempt to solve this problem is being carried on at Princeton University's Forrestal Laboratory; the device is called a *stellerator* because the fusion process is believed to be the one through which the stars (and our sun) release their energy. It is interesting to note that the design of the stellerator has required the application of nearly every one of the fundamental principles discussed in the previous chapters.

14|3 NUCLEAR FORCES

We have seen that in nuclear processes the energy released is around ten million (10^7) times greater than in chemical processes. The release of energy by an atomic system implies that forces in the atom must have done work equivalent to the energy given out. Work equals force times distance. Nuclear distances are of the order of 10^{-15} m, while atomic diameters are 10^5 times greater. For a nuclear force F_n to do, in the distance $s_n = 10^{-15}$ m, an amount of work equal to 10^7 times that done by an electromagnetic force F_e in the distance $s_e = 10^{-10}$ m, we must have

$$F_n \times 10^{-15} = 10^7 \times F_e \times 10^{-10}, \qquad F_n = 10^{12} \times F_e = 100(10^{10}F_e).$$

We must remember that electromagnetic forces vary inversely as the square of the distance between interacting charges. Therefore at a separation of 10^{-15} m the coulomb force between two charges would be 10^{10} times what it is for a separation of 10^{-10} m, that is, $10^{10}F_e$. Thus for a separation of 10^{-15} m the nuclear force between two nucleons must be nearly 100 times the electrostatic Coulomb force.

More information about nuclear forces has been obtained from scattering experiments in which a beam of protons or neutrons is scattered by protons (H^1 nuclei) or by protons and neutrons (nuclei of H^2 or other elements). The stability of the deuteron, which is a proton-neutron combination, and the binding energies of the various known isotopes of all the elements also tell us much about nuclear forces. The following facts about nuclear forces are known.

1. Nuclear forces are *attractive;* they hold the nucleons together in a nucleus.

2. Nuclear forces are very *strong;* they are the strongest forces so far discovered in nature.

3. Nuclear forces are *short-ranged;* they are important only at distances of 10^{-15} m or less.

4. Nuclear forces are *charge-independent;* similar nuclear attractions exist between proton and proton, proton and neutron, neutron and neutron.

5. Nuclear forces are *not central-type forces;* they do not always act along the line joining the interacting particles. Recall that gravitational and electrostatic forces are central, while magnetic forces usually are not.

6. Nuclear forces are *asymmetric.* By this we mean that the force between two nucleons depends on the angles that their joining line makes with the direction of the spin axis of each particle.

7. Nuclear forces *involve an exchange* of a meson particle between the interacting pair of nucleons.

From the above facts we may conclude that the strong nuclear forces represent a new type of force, one not previously described. A fundamental law (perhaps laws) for such forces, analogous to Newton's law of gravitation, Coulomb's law of electrostatic force, and Ampère's law of magnetic force, is being sought. At present it is not clear whether or not this new law may be stated as simply as the others have been; it is felt that more information is needed before it will be possible to make a final statement of the relation between nuclear forces and the various quantities upon which nuclear forces depend. Such a statement could be our *Fundamental Law XVI.*

14|4 THE WEAK INTERACTIONS OF ELEMENTARY PARTICLES

Processes which involve the strong nuclear forces just discussed take place *very* rapidly. Once an α-particle enters a nitrogen nucleus, the ensuing reaction (emission of a proton and formation of an O^{17} nucleus) follows in about 10^{-23} seconds. However, in processes such as radioactive decay with the emission of a beta ray or electron (plus or minus) it is found that the mean decay time is 10^{-10} seconds or longer. This large difference in time suggests that forces much weaker than either those of nuclear attraction or those of electromagnetism are involved in the decay process. It is estimated that these weak forces are not more than 10^{-13} times as strong as are nuclear forces, so they must constitute still another fundamental type of force in the physical world.

All beta-decay processes involve the transformation of one elementary particle into another, and it seems likely that in every case the same fundamental principles are involved. If this is so, we would like to know what fundamental law, new to us, is involved in weak interactions.

14|5 THE FUNDAMENTAL INTERACTIONS

We have seen that at the present time physicists recognize four fundamental types of forces through which particles of matter may interact with one another. In order of increasing strength they are:

1. The gravitational interaction. Call its strength 1.

2. The weak interactions. Relative strength about 10^{25}.

3. The electromagnetic interactions. Relative strength about 10^{36}.

4. The strong interactions. Relative strength about 10^{38}.

More and more information regarding the relative roles played by these interactions is being gathered as increasingly powerful particle accelerators make it possible to create the various elementary particles and to study their decay processes and their interaction with matter. Out of a wealth of experimental information a few empirical rules appear to be emerging, such as the following:

a) *The weaker the interaction the more kinds of particles it applies to.*

b) *A particle subject to one type of interaction is subject to all weaker interactions.*

From the above we see that the gravitational interaction, though weak, is universal since it applies to all particles. If other interactions are absent, as when bodies are massive but uncharged and their distances of separation large, gravitational forces reign supreme.

The weak interaction is experienced by all of the particles listed in the last chapter, but like the gravitational force, it is usually obscured by stronger forces. It is, however, the strongest force felt by the neutrino.

We have seen that the electromagnetic forces are the fundamental ones involved in the structure of atoms and molecules, as well as in the building of macroscopic matter and living cells. Frictional and elastic forces, as well as push and pull, are basically electromagnetic interactions. All charged particles are subject to such forces.

The strong interactions are limited to the baryons and mesons. The chief role of these forces is in the binding together of nucleons to form stable or semi-stable nuclei. However, the role of the somewhat weaker electromagnetic interactions in determining nuclear stability or instability is not insignificant. Due to the Coulomb repulsion between protons, stability in heavy nuclei is favored by the presence of fewer protons than

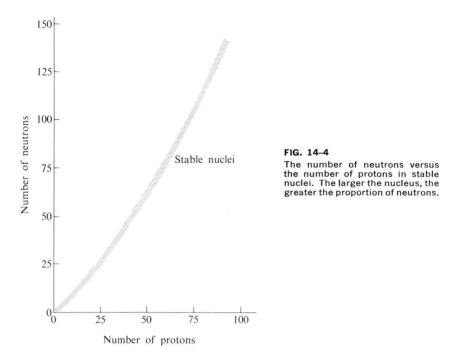

FIG. 14–4
The number of neutrons versus the number of protons in stable nuclei. The larger the nucleus, the greater the proportion of neutrons.

neutrons, as shown in Fig. 14–4. Proton-proton electrostatic repulsion also accounts to a large extent for the increasing instability of the heavy nuclei from the neighborhood of $Z = 84$ (84 protons per nucleus) on to the transuranic elements ($Z > 92$). From the difference between the mass of a nucleus and the computed sum of the masses of its individual protons and neutrons one may calculate theoretically (see Problems 6 and 7 of Chapter 13) how much energy from outside would be required to rip the nucleus apart (a difficult feat to accomplish experimentally). This energy is called the *binding energy* of the nucleus. The relative stability of a nucleus is determined by its binding energy per nucleon, which quantity is plotted against mass number in Fig. 14–5. Note that the binding energies have been plotted downwards so that the lower a point is the more stable the nucleus it represents. Conclusions to be deduced from this curve are left to the reader (see Problem 2 at the end of chapter).

Recall that in connection with the electromagnetic interaction it was convenient to introduce the concept of electromagnetic fields and that such fields were found to travel with the speed of light and to have their energy concentrated in units called *photons*. Similarly a field and corresponding energy unit or quantum have been postulated for each of the other interactions. For the gravitational field the quantum is called the

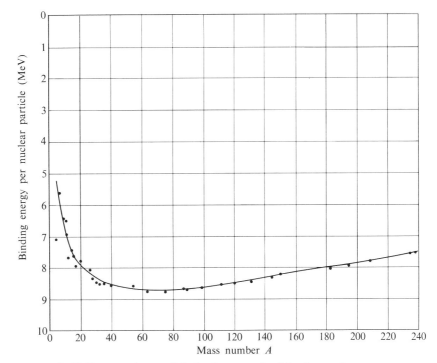

FIG. 14–5. Binding energy (in MeV) per nuclear particle (proton or neutron) for the most stable isotopes of the whole range of mass numbers.

graviton; although there are strong reasons to believe in its existence, it has so far escaped observation.

The quanta of the strong interaction field are believed to be the pi mesons or *pions.* Such a particle was proposed by the Japanese theoretician Yukawa two years before it was discovered experimentally. Just as photons interact readily with electrons, giving energy to, or taking energy from, electrons, so pions interact strongly with nucleons.

We are least certain about the quantum of the weak interaction field, called by some the *W-meson.* It has not been observed and we do not yet understand the weak interaction well enough to know whether such a particle should exist.

14|6 SYMMETRY, INVARIANCE, AND CONSERVATION LAWS

Physicists have become increasingly interested of late in the question of what symmetries nature possesses.

We have seen that certain things may be deduced more quickly by appealing to symmetry considerations rather than by carrying out the detailed application of a physical law to a special situation. For example, in the case of two oppositely charged parallel plates, the electric field is normal to the plates in the region where edge effects may be neglected. This result may be obtained by using Coulomb's law and performing a mathematical integration, but we reach the same conclusion by asking "If the field is *not* normal to the plates, why should one direction be preferred to another?" Obviously the only unique direction is the normal one. Another example of this kind of reasoning occurs in connection with the magnetic field around a long straight wire carrying a current. In the same way the symmetry of a snowflake may be deduced from the assumption of certain conditions and physical laws, but to most people it is an example of nature's symmetry and as such it appeals to our aesthetic sense.

The appeal to symmetry or aesthetic considerations may not always be justified, since it assumes environmental conditions that may not exist. Due to the gravitational, electric, and magnetic fields of the earth, sun, our galactic system, etc., we actually live in a world where perfect symmetry can only be approximated.

There is another kind of symmetry under discussion; it has to do with transformations and conservation laws. For example, we saw that according to the relativity principle, natural laws are invariant under a transformation from one inertial system to another; the results of a physical experiment should be the same in any inertial system. There is thus a symmetry between two inertial systems S and S'. An observer in S finds a clock in S' to be running slow, while an observer in S' says that clocks in S run slow, and so on. What other symmetries of this sort does our world possess?

First of all we should remind ourselves that it is both remarkable and fortunate, as well as a strong argument for belief in God, that we do not live in a chaotic physical world. The laws of physics appear to be independent of space and time. If not, if experiments could not be repeated, we would simply have no laws of nature; we could not invent, we could count on nothing. But due to the symmetry properties of our world there are laws that we can count on under changing conditions and as a consequence of this there are certain things that are *conserved* while other things change. Let us consider some examples.

a | Translation in space

The laws of physics are the same everywhere. Space is homogeneous. The conservation of linear momentum is related to this, for a body moving in empty space that is homogeneous will continue at a uniform velocity, or with constant momentum.

b | Rotation of axes

One direction in space is like another. Space is isotropic and the laws of physics are invariant relative to rotation of our axes of reference. This symmetry property is related to the conservation of angular (rotational) momentum.

c | Displacement in time

The laws of physics do not change from day to day. They are invariant with time. In such a world energy changes within a given system can be accounted for and the total energy of the world remains constant, or is conserved.

d | Time reversal

Suppose that time started to run backward, so that in all of our expressions t would be replaced by $-t$. This would result in all velocities being reversed, but not forces or accelerations (which depend on the time *squared*). Newton's laws and those of electromagnetism would be unaffected. The world would behave as in a moving picture run backward. While some highly *improbable* events would occur, we would observe none that were strictly *impossible*. In other words, only the degradation of energy principle would fail, but this law applies to large groups of particles and need not concern us at the elementary particle level.

e | Charge and matter conjugation

Suppose that we could make a copy world just like our own except that every electron was replaced by a positron, every proton replaced by an antiproton, every neutron by an antineutron, and so on. In this copy world of antimatter Coulomb's and Ampère's laws would hold because both involve the *product* of two charges (or currents) and, with all charges reversed, $\bar{q}_1 \times \bar{q}_2 = (-q_1) \times (-q_2) = q_1 \times q_2$. Newton's laws and the others not relating to charge would be unaffected. Atoms built up out of negative nuclei (containing antineutrons and antiprotons) with orbiting positrons would have the same chemical properties, energy states, and spectra as do their counterparts in our ordinary world, for these properties depend only on the electromagnetic interactions. We may speculate that in some distant part of the universe such a world may actually exist, but, if so, let us hope that we shall never encounter it because, if we did, our world and this world of antimatter would annihilate each other! Perhaps the universe contains equal amounts of matter and antimatter and it is only a matter of chance which kind of matter predominates locally.

343

Related to the symmetry of the world with respect to charge (and matter) is the fact that there are two kinds of charge (and matter), and that charges (and matter) of one sign (and kind) may be separated from charges (and matter) of the opposite sign (and kind), but the net charge (and matter) in a closed system is conserved.

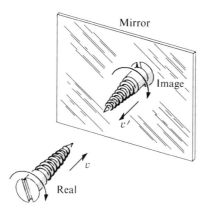

FIG. 14–6
A right-hand screw advancing to the right appears in a mirror as a left-hand screw advancing to the left.

f | Mirror reflection

This is equivalent to reversing one (or all) of our coordinate axes. We would then see a world that would behave as our world would if viewed in a mirror. This amounts to interchanging right-handedness and left-handedness. If a right-hand screw is pointed at a mirror and turned clockwise, as viewed from its rear, the screw will advance toward the mirror (see Fig. 14–6). In the mirror the screw will appear to advance toward the viewer; the viewer will see the screw turning clockwise as viewed from in front, so that he would regard it as turning counterclockwise as viewed from the rear. A screw that advances when turned counterclockwise is a *left-hand* screw. In the mirror world the left hand replaces the right and the right replaces the left.

Would the laws of nature be different in a mirror world than in our own? If so, we could use the laws of nature alone to distinguish right from left. Up until 1956 physicists thought that the laws of nature would be invariant under a mirror-type conjugation. This principle was related to a conservation law called *the conservation of parity*. Parity is defined in terms of the symmetry of the wave functions describing an atomic system or particle, and it has no counterpart in large-scale physics. It was assumed that, in an atomic process, parity does not change and so is conserved. Then in 1956 the theoretical physicists T. D. Lee and C. N. Yang proposed crucial experiments to test the conservation of parity principle. These and other experiments since performed have shown that parity is not conserved in weak interactions and that in such processes nature does distinguish not only between right-handedness and left-handedness, but also between matter and antimatter. However, results indicated that antimatter in a mirror world behaves like matter in our world, or that *the laws of nature are invariant under a simultaneous matter conjugation and mirror reflection*. Recent testing of this postulate suggests that the weak interaction is invariant only under the triple combination of matter conjugation, mirror reflection, and time reversal.

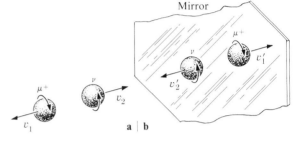

FIG. 14–7
Decay of a π^+ particle into a μ^+ and neutrino results in the situation shown in (a), but not that shown in (b). (b) is the mirror image of (a).

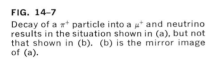

Illustration

When a positive pion (π^+) decays, a positive muon (μ^+) and neutrino (ν) are formed, so that we have

$$\pi^+ \rightarrow \mu^+ + \nu. \tag{14–5}$$

The conservation of momentum (based on Newton's laws) requires that the particles μ^+ and ν be emitted in opposite directions and with opposite spins, as shown in Fig. 14–7(a). We see that both particles are advancing as would left-hand screws. The mirror image of this experiment is shown in Fig. 14–7(b), in which both particles are advancing as would right-hand screws. This mirror-image process has never been found to occur in nature.

If we perform a matter conjugation upon the process of Eq. (14–5), we have

$$\pi^- \rightarrow \mu^- + \bar{\nu}, \tag{14–6}$$

where $\bar{\nu}$ is the antineutrino; π^+ and π^-, as well as μ^+ and μ^-, form particle-antiparticle pairs. The decay process of Eq. (14–6) does occur in nature, but only as shown in Fig. 14–8(a), in which both particles are seen to be advancing as *right-hand* screws. The mirror image of this actual process is shown in Fig. 14–8(b) and it does *not* occur. Hence we see that if we take the process shown in Fig. 14–7(a), which does occur, and perform·upon it both a matter conjugation and a mirror reflection, then we end up with another actual process, one allowed by the laws of nature.

345

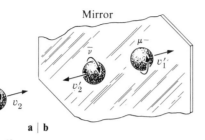

FIG. 14–8
Decay of a π^- particle into a μ^- and anti-neutrino results in the situation shown in (a), but not that shown in (b).

In addition to the conservation laws, or invariance principles, that we have discussed, physicists have assigned other properties, such as *isotopic spin* and *strangeness*, to elementary particles. We need not concern ourselves with the definitions of these concepts but merely note that they obey limited conservation laws. Isotopic spin apparently need be conserved only in processes involving strong interactions, while strangeness is conserved only in connection with strong and electromagnetic reactions. As a general rule one may say that *the weaker the type of interaction involved in a process, the more conservation laws it may violate.* This suggests the possibility that where gravitational forces alone are concerned, particle-number (matter) conservation may not even hold. Thus we have the steady-state theory of the universe, proposed by the British astronomer Fred Hoyle, according to which matter is being continually created and new stars born as old ones die out or move farther apart. This theory does *not* violate the conservation of energy principle because the increase in mass-energy is balanced by a decrease in gravitational energy as the particles newly created are pulled together to form a star.

In summary, the conservation-invariance laws that we have discussed are listed, with those most inviolate at the top and those more often broken at the bottom.

Relativistic invariance
Conservation of energy
Conservation of linear momentum
Conservation of angular momentum
Conservation of charge
Conservation of particles of the electron family
Conservation of particles of the muon family
Conservation of particles of the baryon family
Time reversal invariance
Charge conjugation plus mirror reflection invariance
Conservation of strangeness
Conservation of isotopic spin

14|7 OTHER PROBLEMS

Particle physics is reaching the stage attained by atomic physics when the spectral lines of the various atomic spectra had been carefully measured, but no one knew why hydrogen emitted one type of spectrum, sodium another, mercury a third, and so on. The advent of the Bohr theory made possible a theoretical explanation of the hydrogen spectrum and quantum mechanics has enabled physicists to explain the spectra of all the elements. In particle physics we are confronted with much factual knowledge, but no fundamental theory that explains it all. Such a theory must involve the principle of nuclear forces and that of the weak interaction, but it may also require one or more additional postulates. Perhaps we shall find that just as atoms possess quantized energy states, so matter also has its quantized states represented by the various elementary particles. At heart most physicists believe that the laws of nature are basically simple.

We would like to know why the proton and the electron have the masses they do. Why should we not go further and ask ourselves why the speed of light is 3×10^8 m/sec and not some other value? Why is

$$G = 6.67 \times 10^{-11} \text{ N-m}^2/\text{kg}^2?$$

Thought has been given to the possibility of relationships between the fundamental constants. Any satisfactory theory along this line will have to encompass more than one branch of physics, and so it will need to be a unifying theory, one in which two or more of our fundamental postulates may be combined into a more comprehensive law.

Some physicists have turned from the minute world of the atom to the seemingly infinite one of the universe. In this field of endeavor speculation is rife. Many theories have been proposed to account for the origin of the solar system and the present distribution of stars and interstellar matter in the universe. All such theories must meet the test of explaining known facts, but they cannot be given the thorough testing in the laboratory that we have given the fundamental laws discussed in previous chapters. Cosmic theories will, in the long run, be judged by the same criteria that we apply to other physical theories, such as their rationality, simplicity, fruitfulness, and adaptability.

Are you interested in following man's progress in his search for new fundamental truths? If so, you may choose your reading from an increasing number of excellent books, magazines and newspaper articles written for the educated layman. You are particularly urged to read the *Scientific American*, many of whose articles pertain to the scope of this book.

14|8 GOEDEL'S THEOREM

Kurt Goedel, one of the most brilliant mathematicians of this century, is now at the Institute for Advanced Study in Princeton. In 1931, when he was at the University of Vienna, he published a profound and significant paper on the internal consistency of mathematical systems. Goedel proved what has become a famous theorem, which states among other things that if one builds a mathematical system based on a set of consistent axioms, or postulates, then within the system statements will be discovered such that (1) no exception to these statements is ever found, yet (2) these statements cannot be *proved*, that is, derived from the axioms of the system.

Illustration

A prime number is defined as one that is not divisible by any integer except itself and unity. In number theory it has been found that one can choose any even integer and express it as the sum of two prime numbers, for example, $10 = 7 + 3$, $20 = 17 + 3$, $24 = 19 + 5$, etc. No one has ever found a proof of this theorem as applied to *all* even integers.

We may regard Goedel's theorem as having great import for physics as well as for mathematics. It tells us that if man ever builds what he regards as a complete theory of the physical world, one based on a set of fundamental laws such as we have discussed, then new truths will still be discovered, and these truths will require additional laws for their explanation. In other words, if at some time man wonders whether or not he has discovered *all* of the fundamental laws of nature, he can be sure that he has *not* and that there is at least *one more* fundamental law to look for. The search for new truths should never end!

PROBLEMS

1. What do the answers to Problems 6 and 7 of Chapter 13 indicate with regard to the relationship between stability and mass number for light nuclei?

2. a) Why is it possible to release nuclear energy by fusion of light nuclei and by fission of heavy ones?

   b) What nuclei are the most stable?

   c) Why is iron a poor prospect for a source of nuclear energy?

3. Compute the energy released per kg and per nucleon in the hypothetical fusion of H^1 atoms and neutrons to form helium.

4. Find the so-called *kinetic temperature* T at which the mean thermal energy of a particle $(3kT/2)$ is equal to 10^3 eV. At such a temperature would any of the particles possess the 10^6 eV of energy which is roughly required for a particle to enter into a nuclear reaction?

5. Explain how magnetic forces are not symmetric and usually not central. (See Chapter 8.)

6. Show that Ampère's law would hold in a mirror world. Consider the force between two long straight current-carrying wires as an example.

7. How would the laws of physics be affected by a reversal of all three axes of our inertial system?

8. In the case of a person diving into a pool, how would the laws of physics be affected by a reversal of time, that is, by replacing t with $-t$, or running a movie of the action backward?

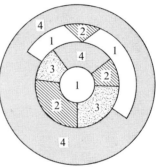

FIG. 14–9
The four-color map problem. The numbers 1, 2, 3, 4, refer to four different colors.

9. Consider as a possible illustration of Goedel's theorem the famous four-color map problem. It is taken as a rule, or axiom, that in a map of several countries, or states, those countries with a common border must be colored differently in order to make them more distinguishable. Under this condition no arrangement of countries has been *found* to exist, or has been imagined to exist, for which more than *four* colors are needed when coloring the map of the countries, but no one has yet *proved* that less than *five* colors will suffice in general. Construct some real or fictitious maps and show that for them four colors are sufficient, as in Fig. 14–9.

GARDNER, MARTIN, *The Annotated Alice*, C. N. Potter, 1960.

References

See also references at end of Chapter 13.

Appendixes

Name	Symbol	Value
Gravitational constant	G	6.67×10^{-11} N-m^2/kg^2
Coulomb's law constant	k_e	8.98×10^9 N-m^2/C^2
Ampère's law constant	k_m	1.00×10^{-7} N-sec^2/C^2
Speed of light in vacuo	c	2.998×10^8 m/sec
Planck's constant	h	6.63×10^{-34} J-sec
Boltzmann's constant	k	1.38×10^{-23} J/°K
Avogadro's number	N_0	6.03×10^{26} molecules/kg-mole
Gas constant	$R = N_0 k$	8.31×10^3 J/kg-mole-°K
Electronic charge unit	e	1.60×10^{-19} C
Faraday's constant	$F = N_0 e$	9.64×10^7 C/kg-mole
Rest mass of electron	m_0	9.11×10^{-31} kg
Ratio of mass of proton to mass of electron	m_p/m_0	1836

Quantity	Symbol	Value
Acceleration due to gravity at		
0° latitude and sea level	g	9.78 m/sec^2
Washington, D.C.	g	9.80 m/sec^2
Speed of sound at standard temperature and pressure	v_s	332 m/sec
Freezing point of water at 1 atmosphere	T_0	273.15°K
Mechanical equivalent of heat	J	4185 J/cal*
Pressure of normal atmosphere	p_0	1.013×10^5 N/m^2
Mass of earth	M_E	5.98×10^{24} kg
Mean radius of earth	R_E	6.37×10^6 m
Mass of moon	M_M	$M_E/81.5$
Mass of sun	M_S	2.0×10^{30} kg
Mean distance to moon	R_{ME}	3.84×10^8 m
Mean distance to sun	R_{SE}	1.5×10^{11} m
Rydberg constant for hydrogen	R_H	1.097×10^7/m
Electron-volt	eV	1.60×10^{-19} J
Angstrom unit	Å	10^{-10} m
Atomic mass unit	amu	1.66×10^{-27} kg

* This is the large calorie, or "kilocalorie."

APPENDIX 3   Powers of Ten and Significant Figures

Calculations in physics frequently involve the multiplication and division of several numbers. The numbers will vary from very small ones, such as 0.000000589, to very large ones, like 603,000,000,000,000,000,000,000,000. Here the smaller number is the wavelength of sodium light, in meters, and the larger number is the number of molecules in 18 kilograms (about 40 pounds) of water. (The large number is Avogadro's number per kilogram-mole.) Since in physics we have to handle numbers ranging so greatly in size, it is necessary to introduce the *power-of-ten notation*. In doing this it is good to adopt the custom of writing one and only one figure in front of the decimal point. Thus we write

$$0.000000589 = 5.89 \times 10^{-7}$$

(the decimal point was moved seven places to the right), and Avogadro's number becomes

$$603 \times 10^{24} = 6.03 \times 10^{26}.$$

This notation has the advantage of shortening such numbers by eliminating the zeros used to mark off the decimal point. How many zeros there are in numbers like these depends upon the choice of units. For example, the wavelength of sodium light may also be expressed as 0.0000589 cm, 0.000589 mm, 5890 Å, etc., so that the zeros here do not play the same role as do the other figures. The zeros referred to are not what are termed *significant figures*. The 5, 8, and 9 above (and the 6, 0, and 3 in Avogadro's number) are to be regarded as having significance within the accuracy of the experimental method by which they were determined, otherwise there would be no point in introducing them; 6.03×10^{26} is regarded as a better estimate of the value of Avogadro's number than, say, 6.04×10^{26} or 6.01×10^{26}. In most cases it is wise for the student to restrict himself in his calculations to three significant figures because this is about what is justified when working with the equipment usually at his disposal. A few measurements, such as weighing with a chemical balance, or measuring potential difference with a potentiometer, may be made more accurately, but these are exceptional. Three-figure accuracy is also about what one obtains from a slide rule and the use of this instrument is a great help when lengthy or repeated calculations are called for.

In multiplying and dividing several numbers the steps are as follows:

1.   Write each factor in the power-of-ten notation.

2.   Multiply and divide the significant figure factors, using three-figure accuracy unless otherwise instructed.

3. Collect powers of ten. Powers of ten in the denominator change sign when brought up into the numerator. Powers of ten in the numerator are added.

Example. Find the value of

$$(1.76 \times 10^{-8}) \cdot (3.55 \times 10^{6}) \cdot (4.4 \times 10^{-2})/(1.60 \times 10^{5}).$$

Solution. In the number 1.60×10^{5} the zero is a significant figure. We find that $1.76 \times 3.55 = 6.2480$, which should be rounded off to 6.25. Then

$$6.25 \times 4.4 = 27.5 \quad \text{and} \quad 27.5/1.6 = 17.2.$$

The powers of ten collect as follows:

$$10^{-8} \times 10^{6} \times 10^{-2} \times 10^{-5} = 10^{-9}.$$

The answer then is 17.2×10^{-9}, which should be expressed as 1.72×10^{-8}, if we adopt the standard form of one significant figure before the decimal.

Definition of the Sine, Cosine, and Tangent   **APPENDIX 4**

Consider the right triangle shown in Fig. A4–1; A and B are its two sides and C its hypotenuse. In physics one frequently finds it useful to employ the ratio of one of the lengths, A, B, C, to one of the others. Hence these ratios have been given names and tables giving their values have been prepared for right triangles of different shapes, that is, right triangles in which the angle θ between B and C has various values from $0°$ up to $90°$ (the greatest possible for a right triangle).

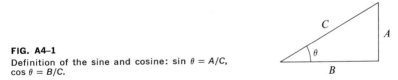

FIG. A4–1
Definition of the sine and cosine: $\sin \theta = A/C$, $\cos \theta = B/C$.

The *sine* of the angle θ is defined as the ratio of the length of the side opposite θ to the length of the hypotenuse. It is conventional to shorten the word *sine* to sin. Then by definition

$$\sin \theta = \frac{A}{C}. \tag{1}$$

In a similar manner the *cosine* of θ, which is shortened to *cos* θ, is defined as the ratio of the length of the side nearest θ to the length of the hypotenuse, or

$$\cos \theta = \frac{B}{C}. \tag{2}$$

Finally, the *tangent* of θ, which is shortened to *tan* θ, is defined as the ratio of the length of the side opposite θ to the length of the side nearest θ, or

$$\tan \theta = \frac{A}{B}. \tag{3}$$

Note that

$$\tan \theta = \frac{A/C}{B/C} = \frac{\sin \theta}{\cos \theta}. \tag{4}$$

To understand what $\sin \theta$ and $\cos \theta$ mean, consider the following points.

1. $\sin \theta$ and $\cos \theta$ are pure ratios. For a given angle, the value of $\sin \theta$, or of $\cos \theta$, is independent of the units used to measure A, B, and C. Of course A, B, and C must be measured in the *same* units.

2. Since the hypotenuse of a right triangle is longer than either side, $\sin \theta$ and $\cos \theta$ must have values between 0 and 1 when θ is between 0° and 90°.

3. From (1) it follows that one may prepare a *table* giving the values of $\sin \theta$ and $\cos \theta$ for various values of θ. Such tables are found in most physics books (see Appendix 5).

4. Since for every value of θ there are corresponding values of $\sin \theta$ and $\cos \theta$, we may think of $\sin \theta$ and $\cos \theta$ as being *functions* of θ. This concept of a function of a variable is an important one. We know that for every value of a variable x there is a corresponding value of x^2, and that x^2 increases as the magnitude of x increases, but at a more rapid rate. We say that x^2 is a *function* of x. Other functions of x are $\sqrt{x}$, $3x^2 + 2x - 5$, $\log x$, $\sin x$, and $\cos x$. So $\sin \theta$ is a function of θ which for small values of θ (say less than 15°) increases nearly proportionally with θ, and which for larger values of θ increases less rapidly than θ and reaches a maximum value of 1 when $\theta = 90°$. $\cos \theta$ is another function of θ which decreases as $\sin \theta$ increases. In fact we should note that

$$\sin^2 \theta + \cos^2 \theta = 1. \tag{5}$$

Can you prove this? (Use the Pythagorean theorem.)

5. Equations (1) and (2), defining $\sin \theta$ and $\cos \theta$, may be rewritten so as to enable us to find either side of a right triangle, given the hypotenuse and the angle θ. Thus in Figure A4–1

$$A = C \sin \theta, \qquad B = C \cos \theta. \tag{6}$$

The values of $\sin \theta$ and $\cos \theta$ may be found from Appendix 5.

Sines and Cosines of Common Angles   **APPENDIX 5**

Angle θ	$\sin \theta$	$\cos \theta$	Angle θ	$\sin \theta$	$\cos \theta$
0°	0.000	1.000	50°	0.766	0.643
5°	0.087	0.996	53°	0.799	0.602
10°	0.174	0.985	55°	0.819	0.574
15°	0.259	0.966	60°	0.866	0.500
20°	0.342	0.940	65°	0.906	0.423
25°	0.423	0.906	70°	0.940	0.342
30°	0.500	0.866	75°	0.966	0.259
35°	0.574	0.819	80°	0.985	0.174
37°	0.602	0.799	85°	0.996	0.087
40°	0.643	0.766	90°	1.000	0.000
45°	0.707	0.707			

Newton's Test of His Law of Gravitation; the Motion of the Moon   **APPENDIX 6**

Let a satellite of mass $m_1 = m$ circle an attracting body of mass $m_2 = M$, the radius of the orbit being R and the period of one revolution T (see Fig. A6–1). The speed v of the satellite may be expressed as the ratio of the distance $(2\pi R)$ around the orbit to the time (T) taken to cover this distance, or $v = 2\pi R/T$.

As shown in Section 4–3, Newton concluded that the satellite must be acted on by a centripetal force whose magnitude is given by

$$F_c = \frac{mv^2}{R} = \frac{m}{R}\left(\frac{2\pi R}{T}\right)^2 = \frac{4\pi^2 mR}{T^2}.$$

Newton next postulated that this force is the gravitational attraction between m and M, or

$$F_c = F_g = \frac{GmM}{R^2}.$$

He equated the two expressions for F_c and obtained

$$\frac{4\pi^2 m R}{T^2} = \frac{GmM}{R^2}.$$

An m may be cancelled on each side and the equation solved for T^2. The result is

$$T^2 = \frac{4\pi^2}{GM} R^3. \tag{1}$$

This is a derivation of Kepler's third law since, for a given planetary system, M is constant and hence T^2 is proportional to R^3. This is a successful *qualitative* accomplishment for Newton's theory.

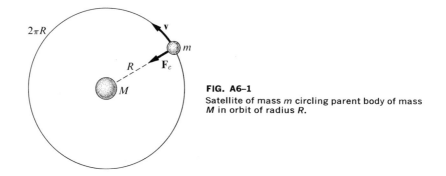

FIG. A6–1
Satellite of mass m circling parent body of mass M in orbit of radius R.

Newton next considered the case where $M = M_E$, the mass of the earth. He equated the weight (mg) of a terrestrial object to the earth's gravitational attraction on it, i.e., he set

$$mg = \frac{GmM_E}{R_E^2},$$

where R_E is the earth's radius, and m is here the mass of the terrestrial object (it cancels out). This equation states that

$$GM_E = gR_E^2. \tag{2}$$

Newton substituted this in Eq. (1), where M now equals M_E, but R is still the radius of the satellite's orbit, and found that

$$T^2 = \frac{4\pi^2}{g} \frac{R^3}{R_E^2}, \qquad T = \frac{2\pi}{R_E} \sqrt{\frac{R^3}{g}}. \tag{3}$$

The radius R_E of the earth was known to be about 4000 miles; R for the moon's orbit was estimated to be 60 times R_E, and g was measured as 32 ft/sec^2 or (32/5280) mi/sec^2. Putting these values in Eq. (3), Newton found for the period of the moon

$$T_M = \frac{2\pi}{4000 \text{ mi}} \sqrt{\frac{(60 \times 4000)^3 \text{ mi}^3}{(32/5280) \text{ mi/sec}^2}} = 2.37 \times 10^6 \text{ sec}.$$

Since there are $24 \times 60 \times 60 = 8.64 \times 10^4$ seconds in a day, the computed value of T_M is equivalent to

$$T_M = \frac{2.37 \times 10^6 \text{ sec}}{8.64 \times 10^4 \text{ sec/day}} = 27.4 \text{ days}.$$

The measured value of T_M is $27\frac{1}{3}$ days; therefore the above computation furnished a remarkable *quantitative* confirmation of Newton's theory.

Some AIP* Abbreviations for Units **APPENDIX 7**

Unit	Abbreviation	Unit	Abbreviation
ampere	A	liter	liter
angstrom	Å	nanosecond	nsec
calorie	cal	megohm	MΩ
centimeter	cm	meter	m
coulomb	C	microcoulomb	μC
day	day	microsecond	μsec
decibel	db	milliampere	mA
electron volt	eV	millimeter	mm
farad (=C/V)	F	newton	N
gram	gm	ohm (=V/A)	Ω
henry (=Wb/A)	H	picofarad	pF
hour	h	second	sec
joule	J	tesla (=weber/m^2)	T
kilocalorie	kcal	volt	V
kilogram	kgm	watt	W
kilohm	kΩ	weber	Wb
kilojoule	kJ	year	yr
kilowatt-hour	kWh		

Note that meg $= 10^6 \times$, kilo $= 10^3 \times$, milli $= 10^{-3} \times$, micro $= 10^{-6} \times$, nano $= 10^{-9} \times$, pico $= 10^{-12} \times$.

* American Institute of Physicists.

Answers to Problems

Chapter 1

1. a) Picture a building with four floors, each floor being a square of sixteen rooms. Opponents alternately choose rooms. Four rooms in a line—horizontally, vertically, or diagonally—wins.

   b) Start with a one-dimensional row of just two squares.

2. All hats are red.

3. Explorer will get the right answer. If native is a liar, he must lie twice.

4. Shorter native tells the truth; taller one lies.

5. $9567 + 1085 = 10{,}652$

6. $8694 \div 63 = 138$

7. Smith

8. The Greens hunt; the Whites drink gin.

9. The families have 5, 4, 3, and 2 children, respectively.

10. a) After a red card play a higher card; after a black card play a lower one.

    b) After odd-numbered card play one of another suit; after even-numbered card play one of the same suit.

Chapter 2

1. a) 1.26×10^{-3} m^2 b) 3.35×10^{-2} m^3 c) 0.477 m/sec

2. kg-m/sec^2 and kg-m^2/sec^2

3. a) m^3/kg-sec^2 b) 6.67×10^{-8} cm^3/gm-sec^2

4. a) 0.4 sec, 1%   b) 39.7 sec

5. Dependence of period on length of arc definitely indicated

6. a) 0.5%   b) 1%   c) 2.5%

7. 20.2 (m/kg)$^{1/2}$, 1.3%

8. Actual error 0.35%, mean deviation 0.2%. Systematic error is indicated.

9. a) $E = vB$ b) $e/m = E^2/2B^2V$

10. a) Yes   b) Restricted

    c) Coefficient of friction, defined as the ratio of force of friction to normal force, depends on surfaces, speed of motion, etc.

Chapter 3

1. a) 9.8, 19.6, 29.4, 39.2, 49.0 m/sec
   b) 4.9, 14.7, 24.5, 34.3, 44.1 m/sec
   c) 4.9, 14.7, 24.5, 34.3, 44.1 m
   d) 4.9, 19.6, 44.1, 78.4, 122.5 m
2. a) 30 N, 12.5 kg   b) Mass   c) Yes
3. a) 1 N, 3.33 kg   b) Yes
4. a) 735 N   b) 733.5 N   c) 184 N
5. 112.5 N to the left and very slightly down
6. 0, 2.5, 5.0, 7.5 m/sec^2
7. $F = mv/\Delta t = 1.45 \times 10^4$ N $= 3.26 \times 10^3$ pounds of force
10. a) 784 N   b) 862 N   c) 706 N
13. a) 100 N by the rope, -100 N by the ground act on boy at left.
    b) -100 N by boy at left, 100 N by boy at right
    c) The 100 N in (a) and the -100 N in (b) are one such pair.
14. 196 N, no
15. a) Always up
    b) Same direction when climbing, opposite going down
    c) Greater going up
16. a) 27 N   b) 5.56 sec
17. a) 10 mi/hr   b) 3.33 mi/hr
18. 476.4 mi/hr
19. a) 0.5 ft/sec   b) 0
20. a) 5×10^5 N   b) 1.12×10^5 pounds of force   c) 15.2 m/sec

Chapter 4

1. a) 3.6×10^{22} N   b) 686 N
2. 3.9×10^{-47} N
3. a) 3.2×10^{-9} N   b) 1.28×10^{-8} m/sec^2
4. a) 1.51×10^3 ft, 10.8 sec   b) 4.20×10^3 ft, 18 sec
5. a) If not, plane's weight will cause it to fall faster than in the circular loop.
   b) 2 mg   c) 4 mg
7. 0.72 AU   b) 1.5 AU
8. Observe its orbit's radius R and its speed v, then use $mv^2/R = GmM/R^2$, M being mass of moon.
9. 8.6×10^2 m/sec, 815 min
10. 54 R_E from earth's center
12. About 7 mi/sec

Chapter 5

1. a) 2940 J   b) 2.94×10^5 J
2. a) 368 W, 0.49 HP   b) 70 W, 0.093 HP
3. 49 J; no
4. $W = 59$ J. Gain in PE $= 49$ J, gain in IE $= 10$ J
5. a) 100 J   b) 400 J

8. a) 6.25   b) 2.5   c) Proportional to weight
9. $M/(M + m)$
10. No
11. 1.5×10^{-2} °C
12. a) 600 J   b) 25%
13. a) Microphone   b) Loudspeaker   c) Photocell
    d) Battery   e) Motor   f) Piston engine
14. a) 400 N/m²   b) 4×10^{-3} atm
15. a) 3:2
    b) 5:2; the most probable distribution becomes more likely, the greater the number of events.

Chapter 6

2. d) Positive x-direction. $a = A, \lambda = 2\pi/k, v = b/k$
3. a) 3×10^9 waves/sec   b) 6×10^{14} waves/sec
4. 15 in all
5. Do, mi, fa, sol, la, do; 11 pleasant two-note and 5 pleasant three-note chords.
8. a) 3 ft   b) Right and left reversed
11. a) Let light enter normal to a small side.
    b) Let light enter normal to hypotenuse.
12. 20 cm
13. a) 13°, 26°, 41°   b) 19°, 41°, 82°
14. a) Red   b) Blue   c) Black
15. Length must equal $m \lambda/2, m = 1, 2, 3, \ldots$

Chapter 7

1. a) 6.25×10^9 b) 2×10^{-8} C
2. 64×10^{-5} N   b) 64×10^{-5} N
3. a) Balls now have like charges   b) $+4 \times 10^{-9}$ C
   c) 27×10^{-5} N, 9×10^{-5} N
4. 4500 N/C, directed in second quadrant
7. a) 5×10^3 V/m   b) 8×10^{-16} N up
8. a) 6.4×10^{-18} J   b) 6.4×10^{-18} J   c) 3.78×10^6 m/sec
11. a) 400 V   b) 2×10^4 V/m   c) No
    d) Yes, from work of separation
12. a) 100 V   b) 5×10^3 V/m   c) Yes
    d) No, but voltage supply gained energy.
13. 2000 W
14. a) $-q_2$ b) $q_1 + q_2$

Chapter 8

1. 1.28 A up
2. 8×10^{-5} A
5. a) Around the magnetic poles   b) Near the magnetic equator

6. 1.5×10^{-4}, 6×10^{-4}, 13.5×10^{-4}, 24×10^{-4} N, or 1.53×10^{-2}, 6.12×10^{-2}, 13.8×10^{-2}, 24.5×10^{-2} gm-wt
7. 2.1×10^{-3} N toward center of square
8. Attract
9. a) 1.2 N/A-m   b) 1.92×10^{-12} N
10. 1 A
11. 10^4 V/m, 10^{-3} N/A-m, 10^7 m/sec
12. a) 5×10^7 C/kg   b) 6.4×10^{-27} kg   c) 10^6 V

Chapter 9

1. a) 300 J   b) 1200 J   c) Yes
2. a) 300 J   b) 1440 J   c) No
3. 5, 5 ohms for first conductor; 5, 6 ohms for second, which does not obey either Ohm's or Joule's law.
4. a) 2 A, 5 A, 7 A   b) 220 W, 550 W; lower resistance
5. a) 4 V   b) 12 W   c) 0.3 N   d) 12 W
6. a) 5×10^{-2} V   b) 2.5×10^{-2} V
7. a) 2.5×10^{-3} A, 2.5×10^{-4} C   b) 1.25×10^{-3} A, 2.5×10^{-4} C
8. $\pi l^2 n B$
9. a) Counterclockwise
10. a) 2200 V, 0.5 A   b) 2200 V, 0.45 A
11. 4.67×10^{-6} J/m^3
12. Neither gain nor lose energy
13. Gains

Chapter 10

1. a) 50 sec, 62.5 sec   b) 80 ft
2. a) 2.6×10^8 m/sec   b) 400 lb   c) 33 yr
3. $m_0 v/\sqrt{1 - v^2/c^2}$
4. a) 5.1×10^4 eV   b) 5.1×10^5 eV   c) 4.6×10^6 eV
5. a) 9.42×10^7 eV   b) 9.42×10^8 eV   c) 8.5×10^9 eV
6. a) 2.56×10^3 eV   b) 7.86×10^4 eV   c) 6.60×10^5 eV
   d) 3.1×10^6 eV
7. 6.1×10^{-12} kg
8. 4.4×10^9 kg/sec
9. 19.5 yr
10. 319 days
11. a) 2280 kg   b) 128,000 kg

Chapter 11

1. a) 2   b) 1
   c) 12,400, 6200, 4133, 3100, 2480, 1240, 124, 1.24, 0.0124 Å, respectively.
2. 6170°K
3. a) 5.06   b) 2.38
4. a) $(hc/\lambda) - E$ b) $(hc/\lambda) + E$

6. a) 3.90×10^{-15} J-sec/C b) 6.24×10^{-34} J-sec c) 1.6 eV
7. 1215 Å, 6562 Å, 4861 Å, 4340 Å
8. V eV, photoelectric effect
9. a) 0.248 Å b) 0.05 Å

Chapter 12

1. 3.85×10^{-5}
2. b) 1.47×10^{-10} m
3. 2.76×10^{-10} m radius
4. 18 states for $n = 3$, 32 for $n = 4$
5. $n = 4$, $l = 3$, $m_l = 3, 2, 1, 0, -1, -2, -3$, $m_s = \frac{1}{2}, -\frac{1}{2}$
6. It has 10 electrons in $n = 3$, $l = 2$ states.
7. a) K^+, I^- b) Cu^+, Cl^- c) Cu^{2+}, Cl^-
   d) Mg^{2+}, O^{2-} e) Al^{3+}, Cl^- f) Al^{3+}, O^{2-}
8. NaCl ionizes more readily and is less stable than H_2O.

Chapter 13

1. 15 gm of water evaporated
2. 5.33 kg
3. 43 gm
4. 4.17 eV, 0.35 eV/nucleon, 0.095 eV/nucleon
5. a) Alpha particle b) Neutron c) Proton
   d) Proton e) Neutron f) Neutron
6. 28.2 MeV, 7.05 MeV/nucleon
7. a) 7.66 MeV b) 7.90 MeV
8. 17.5 MeV
9. 0.78 MeV
10. Final nuclei are C^{13}, O^{16}, O^{17}, respectively.
11. a) 16 days b) 53 days c) 80 days
12. a) 36% b) 59%
13. a) Matter b) Proton, electron, neutrino, 2 antineutrinos c) $+1$

Chapter 14

1. The larger the mass number, the greater the stability
2. a) Binding energy per nucleon is greatest for nuclei of medium mass.
   b) Ones with mass number near 60
   c) It has almost maximum binding energy per nucleon.
3. 1.88×10^8 kWh/kg, 7.05 MeV/nucleon
4. 7.7×10^6 °K, yes
7. Same as mirror reflection
8. Improbable, but not impossible, events would appear; the degradation of energy law alone would be disobeyed.

Index

ABCDE6987